ST(P)
Technology
Today Series

Engineering Design for TEC Level III

ST(P) TECHNOLOGY TODAY SERIES

A full list of titles in the series is available from the publishers on request post free.

The series covers mathematics and statistics for Mechanical and Electrical Engineering; Physical, Applied and Mechanical Engineering Science; Engineering Instrumentation and Control; Engineering Design; Electronics; Building Construction.

Engineering Design for TEC Level III

M D BROOKS B Ed T Eng A M I Prod E AMIED Cert Ed
and
D OLDHAM C Eng M I Mech E Cert Ed
Rotherham College of Arts and Technology

Stanley Thornes (Publishers) Ltd

First published in 1981 by Stanley Thornes (Publishers) Ltd, Old Station Drive, Leckhampton, CHELTENHAM GL53 0DN, England.

Reprinted 1988

British Library Cataloguing in Publication Data

Brooks, M.D.
    Engineering design — TEC level III.
    1. Engineering design
    I. Title   II. Oldham, D.
    620'.004202462   TA174

ISBN 0-85950-303-8

Typeset by Tech-Set, Gateshead, Tyne & Wear
Printed in Great Britain at The Bath Press, Avon

In memory of the late W L Kendall, Senior Lecturer in the Department of Mechanical and Production Engineering, Doncaster Metropolitan Institute of Higher Education

Contents

Preface

This book is intended for students studying for the TEC Certificate and Diploma in Mechanical and Production Engineering, (TEC programme A5).

It covers the general objectives, with liberal illustrations, for Engineering Design III, by describing the factors influencing the choice of material in a component or assembly; comparing the basic manufacturing process for components, with regard to cost and production problems; and analysing ergonomic and safety aspects involved in design.

The student is introduced to the design of simple mechanisms, and a case study in design is examined comprehensively to provide a formative approach. A particular feature of the book are the assignments specifications, written to the recommendations of the British Standards Institution's publication, 'Guide to the Preparation of Specifications, PD6112'. These specifications will aid the student in the formulation of designs for the assignments.

Questions have been set on each chapter to test the students knowledge of the topic areas.

As planned at present, there will be additional books to cover the following TEC units:

Engineering Drawing I
Engineering Drawing II

M D BROOKS
D OLDHAM   Rotherham 1981.

Acknowledgements

We gratefully acknowledge the help given by our colleagues at Rotherham College of Arts and Technology in the areas of their expertise, and by Mr G D Redford who checked our original script in detail.

Our special thanks are also due to the following for permission to reproduce copyright material or for allowing us to include diagrams and details of their processes and products:

Alexander Machinery Ltd
Avery-Denison Ltd
British Standards Institution
Coats Machine Tool Co
The Colchester Lathe Co Ltd
Delta Extruded Metals Co Ltd
Denis Leader (Machinery) Ltd
Department of Industry
Doncasters, Sheffield, Ltd
Edu-Tex Ltd
J H Fenner & Co Ltd
HMSO
A Kinghorn & Co Ltd, Todmorden
J Muscroft Engineering Ltd
National Association of Drop Forgers and Stampers
Neill Tools Ltd, Sheffield
Plastic Coatings Ltd
Sciaky Ltd
Sigma Ltd
Stanley Tools Ltd
Talbot Motor Co Ltd
Victoria Machine Tool Company
Walton of Radcliffe, Manchester
Zinc Alloy Die Casters Association
Zinc Development Association

Material from BS 5304: 1975 is reproduced by kind permission of the Health and Safety Executive, and the British Standards Institution from whom complete copies can be obtained. Material from the booklet *Controlling Corrosion* is used in Chapter 1 and is reproduced by kind permission of the Department of Industry.

We would also like to express our appreciation for the friendly cooperation and helpful advice given to us by the publisher.

Thanks are also due to Mrs Pat Smith for the excellent typing of the manuscript and to the College Librarian Miss B C Lim, ALA for her help.

Finally we would like to add a word of thanks to our wives, Patricia and Maureen, for their patience, help and encouragement during the preparation of this book.

M D Brooks
D Oldham

MATERIALS

The Factors Influencing the Choice of Material

After reading this chapter you should be able to:

★ describe the factors influencing the choice of material in a component or assembly (G);

★ indicate how the choice of material can be affected by various factors such as fatigue properties, anti-corrosive properties, costs, etc. (S).

(G) = general TEC objective
(S) = specific TEC objective

INTRODUCTION

We tend to take for granted the materials from which articles, in everyday use, are manufactured. Seldom do we question why an article has been made from a particular material. The range of materials used to make products such as toothbrushes, electronic calculators, motor cars, etc. is indeed vast.

A designer needs to consider many factors, which may influence the choice of material, for a component which is to perform a particular function. This chapter endeavours to help the student to be aware of the properties and cost factors of materials and the affect these have upon manufacturing methods, and to show how these factors all come under the consideration of the designer in the course of designing a product.

A designer will use his engineering experience and study, keeping himself abreast of new developments, to help select the correct material. The selection will be governed by the conditions of service, which may vary widely from product to product. Every component of an assembly must be designed to a minimum mass, but consideration must be given to the safety aspect in order that reduced mass, with the consequential reduced cost, will not make the product dangerous in its use.

The designer will use his knowledge and experience when undertaking calculations to determine the strength required by the component in service. The questions he will ask himself are shown in Fig. 1.1.

Let us now consider some of the factors influencing the choice of material in greater detail.

Component shape or size
simple or complex? Large or small?

Is the mass of the component
critical?
Is a high-density or low-density
material more satisfactory?

Does the component need to
have good thermal or
electrical conductivity
properties?

Is the component subject
to temperature levels or
variations which are
within the critical range
of the properties of some
materials?

Does the component
need to be hardened
to prevent wear or
indentation?
Is it subject to
impact or fatigue?

If the temperature of
the component will vary
during its use, will
expansion or contraction
present problems?

Does the component
need to have
anti-corrosive properties?

What mechanical
strength is required
by the component?
Does it need to have high
tensile, compressive or shear
strength?
Would a plastic material
be most suitable?

Dimensional accuracy of component
±2mm or ±0.02mm?

What quantities are
to be produced?
This will have a bearing
upon the method of
manufacture, i.e. is it to be
cast, extruded or moulded in
plastic? This in turn will
affect the choice of material
and consequently the cost of
the product.

Fig. 1.1   The questions a designer must ask himself whilst designing.

PHYSICAL PROPERTIES ——————————————————

Density

If we pick up blocks of equal volumes of lead, steel and aluminium it is obvious that their weights, and consequently their masses, are different. The heaviest, lead, is denser than steel, which in turn is denser than aluminium.

Density is defined as the mass per unit volume, and can be calculated from

$$\text{Density} = \frac{\text{Mass}}{\text{Volume}} \text{kg/m}^3$$

The density of a material is an important factor where the weight, and thus the mass, of a component is critical. Aluminium alloys are used in engine parts and extensively in aircraft manufacture where their lightness and medium strength have energy-saving advantages. Plastic, another low-density material, is used in many domestic appliances, where it has the additional advantages of being corrosion-resistant, colourful and attractive in appearance and pleasant to handle. Some of the uses of low-density materials are shown in Fig. 1.2.

Conductivity

If we heat two rods of equal thickness and length, one of copper and one of glass, and hold one in each hand, we will find the copper rod will soon become too hot to hold. The glass rod will be warm, but can be held much longer. Both rods will have the ability to conduct heat, but at different rates. This is the property of thermal conductivity.

In general, metals have high thermal conductivity, the best conductors being silver, copper and aluminium. Copper is used for soldering irons so that the heat is conducted easily to maintain a heat source at the soldered joint. Aluminium is used in saucepans so that heat can be conducted quickly to the contents.

Materials with low thermal conductivity may be used to advantage where the conduction of heat needs to be reduced. Examples of these are plastic handles for cooking utensils. Air is also a poor conductor of heat and is used for insulation in double glazing and in the cavity wall in house structures. To give additional insulation the cavity is often filled with plastic granules or foam (see Fig. 1.3).

All metals and certain non-metals, such as carbon, are also good conductors of electricity – they have high electrical conductivity. The physical state of the material and the temperature will influence its level of conductivity. Silver, copper and aluminium have high

Aluminium alloys and plastics
used for electric drill
bodies and switches

Aluminium pressings and
plastic for handles and
housings on electric
kettles

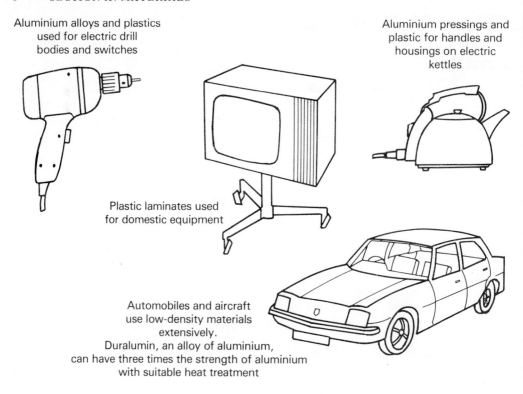

Plastic laminates used
for domestic equipment

Automobiles and aircraft
use low-density materials
extensively.
Duralumin, an alloy of aluminium,
can have three times the strength of aluminium
with suitable heat treatment

Aluminium and its alloys are used for crankcases,
cylinder blocks and heads, gear castings, inlet
manifolds, pistons and connecting rods

Magnesium is lighter than aluminium and has a
very high strength/weight ratio. Magnesium
alloys are used for crankcases and
transmission casings

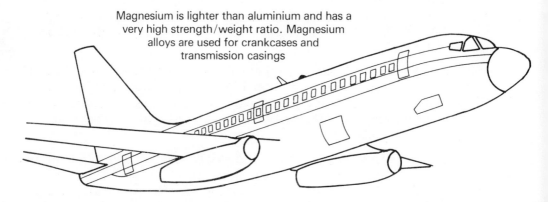

Plastic mouldings are used for knobs, distributor caps, fuse blocks,
instrument panels, window trims and for electrical parts. Man-made
fibres are used in the manufacture of upholstery. Use is also made of
PVC for seals and gaiters, nylon for fan blades and PTFE for
unlubricated bushes

Fig. 1.2   Applications for low-density materials.

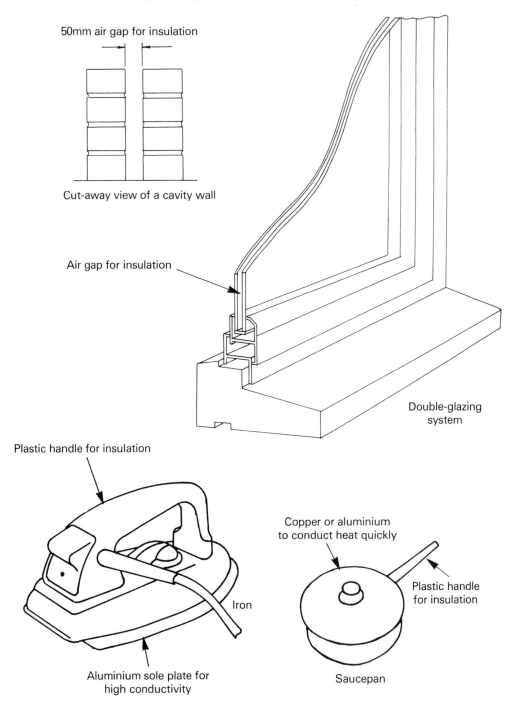

50mm air gap for insulation

Cut-away view of a cavity wall

Air gap for insulation

Double-glazing system

Plastic handle for insulation

Copper or aluminium to conduct heat quickly

Plastic handle for insulation

Iron

Aluminium sole plate for high conductivity

Saucepan

Fig. 1.3 Examples of materials with high and low thermal conductivity.

electrical conductivity, and the choice between them is often determined by strength and cost. Aluminium is often used in place of copper because it is cheaper and has a lower density, giving a consequent weight advantage. This is particularly useful in applications such as overhead power cables.

An insulator is a material with a low electrical conductivity which can be used to separate conductors from each other. Good insulating materials are glass, air, rubber, PVC and ceramics (see Fig. 1.4).

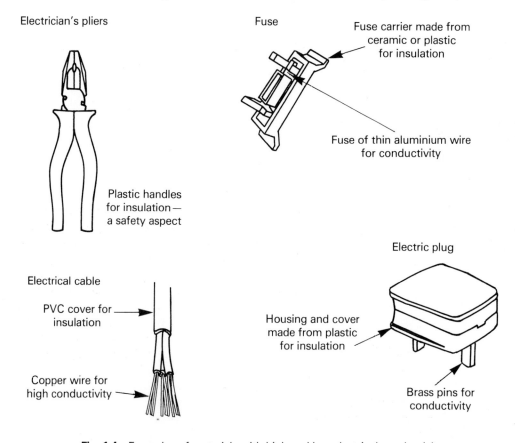

Electrician's pliers

Fuse

Fuse carrier made from ceramic or plastic for insulation

Fuse of thin aluminium wire for conductivity

Plastic handles for insulation — a safety aspect

Electrical cable

PVC cover for insulation

Copper wire for high conductivity

Electric plug

Housing and cover made from plastic for insulation

Brass pins for conductivity

Fig. 1.4   Examples of materials with high and low electrical conductivity.

Melting Point

As heat energy is added to a substance its temperature will be raised by an amount dependent upon the type and mass of the material. If equal amounts of heat energy are added to different substances, of equal mass, the increases in temperature will be different. For example, if equal amounts of heat energy are added to equal masses of tin and steel, the increase in temperature of the tin will be twice that of steel.

Substances which are solid at normal temperatures may become liquid at higher temperatures. The temperature at which this takes place is the melting point of the substance. In general, strong metals have higher melting points than weaker metals (see Table 1.1).

Table 1.1 *Melting points of some metals*

Metal	°C
Tungsten	3370
Molybdenum	2620
Chromium	1830
Nickel	1455
Carbon steel (average)	1370
Cast iron (average)	1180
Copper	1083
Gold	1063
Silver	961
Brass (60 Cu/40 Zn)	900
Aluminium	660
Magnesium	650
Zinc	420
Lead	327
Tin	232

Under load at high temperatures, metals tend to resemble plastics in the manner in which they deform, and consequently many applications need to be considered at the design stage. High-temperature applications such as superchargers, turbines and aircraft engines require materials which will operate under conditions of extreme temperature and stress. Materials used for these applications are nickel-chromium and stainless steels and titanium. For high-temperature applications such as hot working dies and moulds, steels will be selected which also have resistance to abrasion, pressure, shock and fatigue. Tungsten, molybdenum and chromium steels may be used.

Coefficient of Expansion

We have seen in the previous section how temperature increases as heat energy is added to a material. At the same time it will expand in all directions. Similarly if heat energy is withdrawn from a material the temperature will fall and the material will contract. These affects will have considerable importance where the design may be subject to temperature variations.

In engineering practice, expansion and contraction are considered for:
   (i) the linear dimensions of a solid — *linear expansion* or *contraction,*
   (ii) the volume of a liquid or solid — *volumetric* or *cubic expansion* or *contraction.*

Substances will expand or contract at different rates. The amount of expansion or contraction will depend upon the type of material, the change in temperature, and size or volume of the substance.

Linear Expansion

The linear expansion of a liquid or solid can be calculated from:

Expansion or contraction (x) = Coefficient of linear expansion (α)

$\times$ Original length (l)

$\times$ Temperature change (t)

i.e. $x = \alpha l t$

The coefficient of linear expansion is the increase in length per unit length for every one degree rise in temperature.

Some typical values of the coefficient of linear expansion for some metals are shown in Table 1.2.

Table 1.2   *Coefficients of linear expansion (α) of some metals*

Metal	(α) per $°C$
Aluminium	23.7×10^{-6}
Brass	19.5×10^{-6}
Copper	16.7×10^{-6}
Iron (cast)	10.2×10^{-6}
Lead	29.0×10^{-6}
Nickel	12.8×10^{-6}
Silver	19.0×10^{-6}
Steel	11.2×10^{-6}
Tin	21.4×10^{-6}
Zinc	28.2×10^{-6}

Example 1 A cast iron steam pipe is 20 m long at 20 °C. Calculate the expansion when the pipe is carrying steam at a temperature of 165 °C.

Solution   Expansion = $\alpha l t$

= $10.2 \times 10^{-6} \times 20\,\text{m} \times (165°C - 20°C)$

= $0.030\,\text{m}$

$\therefore$ Expansion = 30 mm

Example 2 A brass ball is heated to a temperature of 350°C and its diameter measures 92.50 mm. It is then placed over a slot which is 92.25 mm wide. At what temperature will the ball fall into the slot?

Solution   For the ball to fall into the slot, the ball must contract from 92.50 mm to 92.25 mm diameter, i.e. 0.25 mm.

From $x = \alpha l t$, $$t = \frac{x}{\alpha l}$$

$$\therefore \text{ Temperature change } (t) = \frac{0.25}{19.5 \times 10^{-6} \times 92.50}$$

$$= 138.6\,°\text{C}$$

$\therefore$ Temperature at which the ball will fall into the slot

$$= 350°\text{C} - 138.6°\text{C} = 211.4°\text{C}, \quad \text{say } 211°\text{C}.$$

Example 3 A copper bar 50 mm long at room temperature, 20°C, is to be replaced by one of aluminium in an electrical switching mechanism. If the mechanism is to be actuated at a temperature of 75 °C, determine the original length of aluminium bar.

Solution   Expansion of copper at 75 °C $= \alpha l t$

$$= 16.7 \times 10^{-6} \times 50 \times (75 - 20)$$

$$= 0.045 \text{ mm}$$

$\therefore$ Final operating length of copper bar $= 50 + 0.045$

$$= 50.045 \text{ mm}$$

This is also the same length of the expanded aluminium bar.

$\therefore$ Contraction of aluminium bar at 20 °C $= \alpha l t$

$$= 23.7 \times 10^{-6} \times 50.045$$
$$\times (75 - 20)$$

$$= 0.065 \text{ mm}$$

$\therefore$ Original length of aluminium bar $= 50.045 - 0.065$

$$= 49.98 \text{ mm}$$

Practical examples where a designer will need to make allowance for expansion or use it to advantage are:

(i) at the ends of large metal bridges where the difference in winter and summer temperatures can give considerable changes in the overall length of the bridge. A structure such as this is usually anchored at one end and supported on rollers at the other end to allow for these variations.

(ii) in a bimetallic strip, which consists of two strips of different materials which are brazed together. This can be used as a device for thermostatically controlling an operating switch which will open when a predetermined temperature has been reached. The two metals have different rates of expansion and will bend, when the temperature is increased, and break the contacts (see Fig. 1.5(a)).

(iii) in a curved bimetallic strip system which may be used for a domestic central heating thermostat or automatic fire-alarm. Any change in temperature causes the free end to move, thus turning the pointer via a simple chain mechanism (see Fig. 1.5(b)).

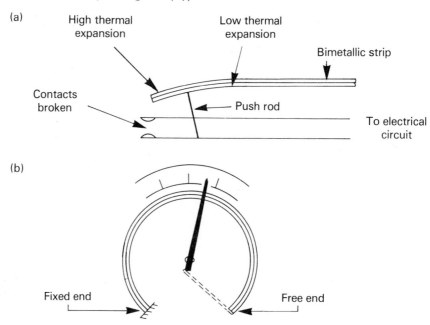

(a)

High thermal expansion   Low thermal expansion

Bimetallic strip

Contacts broken

Push rod

To electrical circuit

(b)

Fixed end   Free end

Fig. 1.5 Practical uses of linear expansion: (a) a simple bimetallic swich;   (b) a domestic-type bimetallic thermometer.   *(By courtesy of Edu-Tex Ltd.)*

Volumetric Expansion

The volumetric expansion of a liquid or solid can be calculated from:

Expansion or contraction (*v*) in volume = Coefficient of volumetric expansion (γ)

X Original volume (V)

X Temperature change (t)

i.e. $v = \gamma V t$

The coefficient of volumetric expansion is the increase in volume per unit volume for every one degree rise in temperature.

For solids $\gamma = 3\alpha$.

Example 1 A cube of aluminium of side 25 mm is subject to a temperature rise of 115 °C. Determine its percentage increase in volume.

Solution   Coefficient of linear expansion for aluminium $(\alpha) = 23.7 \times 10^{-6}/°C$.

Coefficient of volumetric expansion for aluminium

$$(\gamma) = 3\alpha = 3 \times 23.7 \times 10^{-6}$$

$$= 71.1 \times 10^{-6}/°C$$

Original volume of aluminium $= 25 \times 25 \times 25 \text{ mm}^3 = 15\,625 \text{ mm}^3$

Expansion $(v) = \gamma Vt$

$$= 71.1 \times 10^{-6} \times 15\,625 \text{ mm}^3 \times 115 °C$$

$$= 128 \text{ mm}^3$$

$$\therefore \% \text{ increase in volume} = \frac{\text{Increase in volume}}{\text{Original volume}} \times 100$$

$$= \frac{128}{15\,625} \times 100 = 0.82\%$$

Example 2 A thermometer of 1 mm bore contains 4600 mm³ of mercury at 20 °C. Determine the increase in height of the mercury when the temperature has reached 45 °C. The coefficient of volumetric expansion of mercury is $18 \times 10^{-5}/°C$.

Solution   Expansion in volume $(v) = \gamma Vt$

$$= 18 \times 10^{-5} \times 4600 \times (45 - 20)$$

$$= 20.7 \text{ mm}^3$$

Volume of a cylinder $= \pi r^2 h$

$$\therefore h = \frac{v}{\pi r^2} = \frac{20.7}{\pi \times 0.5^2} = 26.4 \text{ mm}$$

$$\therefore \text{ Height of mercury} = 26.4 \text{ mm}$$

Practical examples of the use of volumetric expansion are:

(i) the mercury-in-glass thermometer where an increase in temperature causes the mercury to expand up the capillary tube (see Fig. 1.6(a));

(ii) the mercury-in-steel thermometer where the expansion of mercury causes an increase in pressure which is measured by a suitably calibrated pressure gauge (see Fig. 1.6(b)).

(a)

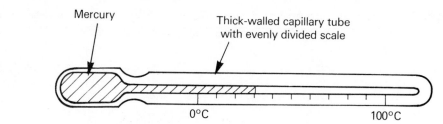

(b)

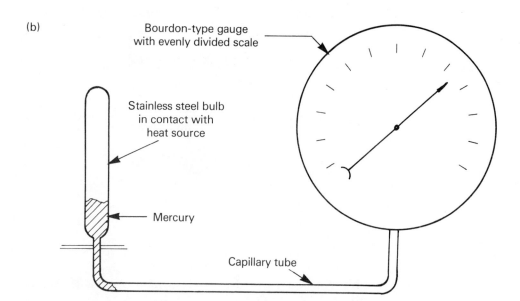

Fig. 1.6 Practical use of volumetric expansion: (a) a mercury-in-glass thermometer; (b) a mercury-in-steel thermometer. *(By courtesy of Edu-Tex Ltd.)*

MECHANICAL PROPERTIES ———————

Strength of Materials

Strength is the ability of a material to resist fracture when subjected to a force. All materials will break if a force of sufficient size is applied to them. If a component is subjected to an external load, an internal force is set up in the component which enables it to resist the loading. When the internal force is uniformly distributed over a cross-section, the *stress* is the internal force per unit area, and is calculated from

$$\text{Stress} = \frac{\text{Applied force}}{\text{Cross-sectional area}}$$

The basic unit of stress is the newton per square metre (N/m^2), but for practical purposes this is too small and the units meganewton per square metre (MN/m^2) are used.

N.B. It is useful to note when undertaking calculations, that
$$1\,N/mm^2 = 1\,MN/m^2$$

There are three fundamental types of stress to be considered:

   (i) tensile stress, which is set up by forces trying to pull the material apart (see Fig. 1.7(a));

  (ii) compressive stress, which results from forces trying to compress the material (see Fig. 1.7(b));

 (iii) shear stress, which is produced by forces which tend to try to make one portion of the material slide over another portion (see Fig. 1.7(c)).

Example 1 A wire of diameter 5 mm has a tensile force of 120 N applied to it. Calculate the stress produced in the wire.

Solution

$$\text{Force in wire} = 120\,N$$

$$\text{Cross-sectional area of wire} = \frac{\pi d^2}{4} = \frac{\pi \times 5^2}{4} = 19.64\,mm^2$$

$$\text{Stress} = \frac{\text{Force}}{\text{Area}} = \frac{120\,N}{19.64\,mm^2} = 6.11\,N/mm^2 \quad \text{but}$$

$$1\,N/mm^2 = 1\,MN/m^2$$

$$\therefore \text{Tensile stress} = 6.11\,MN/m^2$$

Example 2 A square block of material of side 8.25 mm is compressed such that a stress of 94 MN/m² is induced in it. Calculate the compressive force which is being applied.

Solution

$$\text{Cross-sectional area of material} = 8.25 \times 8.25 = 68.06\,mm^2$$

$$\text{Stress} = 94\,MN/m^2 = 94\,N/mm^2$$

$$\text{From Stress} = \frac{\text{Force}}{\text{Area}}, \quad \text{Force} = \text{Stress} \times \text{Area}$$

$$\therefore \text{Force} = 94\,N/mm^2 \times 68.06\,mm^2$$

$$= 6400\,N$$

$$\therefore \text{Force applied} = 6.4\,kN$$

(a)

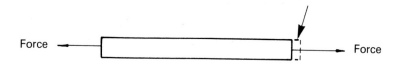

Length tends to increase

Force

Force

(b)

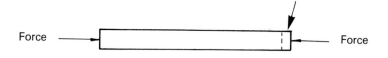

Length tends to decrease

Force

Force

(c)

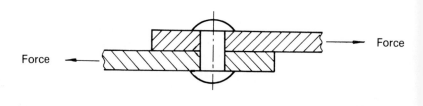

Force

Force

Rivet shearing

Fig. 1.7   Fundamental stresses produced in components: (a) tensile stress;   (b) compressive stress;   (c) shear stress (in a rivet).

Example 3 A press tool is required to pierce a slot in a steel strip 5 mm thick. The slot is 15 mm wide and 45 mm overall length and has semi-circular ends (see Fig. 1.8). If the shear strength of the material is 320 MN/m² calculate:

    (i) the force required to punch out the slot;

    (ii) the magnitude and type of stress set up in the punch.

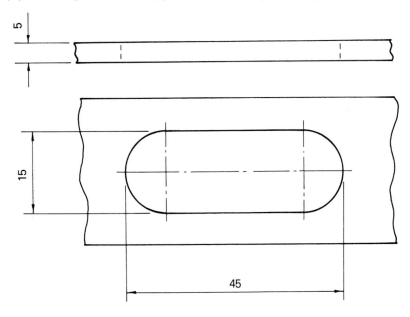

Fig. 1.8 The slot required in Example 3.

Solution

(i) Area of slot resisting shear = Perimeter of slot
$$\times \text{Thickness of material}$$

$$\text{Perimeter of slot} = \cancel{2} \times \frac{\pi d}{\cancel{2}} + (2 \times 30)\ \text{mm}$$

$$= 107.13\ \text{mm}$$

$$\therefore \text{Area resisting shear} = 107.13 \times 5 = 535.65\ \text{mm}^2$$

$$\text{Force} = \text{Stress} \times \text{Area}$$

$$= 320\ \text{N/mm}^2 \times 535.65\ \text{mm}^2$$

$$= 171.4\ \text{kN}$$

(ii) The type of stress set up in the punch is compressive.

$$\text{Cross-sectional area of the punch} = \text{Area of slot}$$

$$\text{Area of slot} = \cancel{2} \times \frac{\pi d^2}{\cancel{4}} + (30 \times 15) = 626.7\ \text{mm}^2$$

$$\text{Force applied by the punch} = \text{Force required to punch}$$
$$\text{out the slot, from (i)}$$

$$\therefore \text{ Force in punch } = 171.4 \text{ kN}$$

$$\therefore \text{ Stress in punch } = \frac{\text{Force}}{\text{Area}} = \frac{171\,400\,\text{N}}{626.7\,\text{mm}^2} = 274\,\text{N/mm}^2$$

$$= 274\,\text{MN/m}^2$$

$$\therefore \text{ Compressive stress in the punch } = 274\,\text{MN/m}^2$$

Example 4 A double lap joint consists of 6 rivets each 8 mm diameter. Determine the maximum shear force which can be supported by the joint if the stress in the rivet material must not exceed $40\,\text{MN/m}^2$.

Solution   Fig. 1.9 shows the double lap joint.

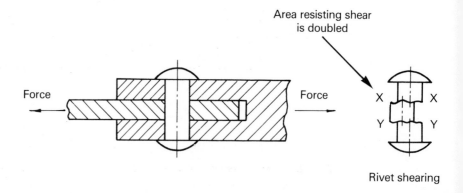

Fig. 1.9   The double lap joint in Example 1.4.

The joint shown in Fig. 1.7(c) is known as a single lap joint and the rivets are said to be in single shear.

In this example the rivets are in double shear since shearing can take place in both planes XX and YY. This type of joint is twice as strong as a single lap joint as the area resisting shear is doubled.

$$\text{Area resisting shearing force in one rivet} = \frac{\pi d^2}{4} \times 2 \quad (\text{double shear})$$

$$= \frac{\pi \times 8^2}{4} \times 2$$

$$= 100.54\,\text{mm}^2$$

Total area resisting shearing force $=\ 100.54\,\text{mm}^2 \times$ Number of rivets

$$=\ 100.54 \times 6$$

$$=\ 603.24\,\text{mm}^2$$

Force $=$ Stress $\times$ Area

$$=\ 40\,\text{N/mm}^2 \times 603.24\,\text{mm}^2$$

$$=\ 24.13\,\text{kN}$$

$\therefore$ Maximum shear force $=\ 24.13\,\text{kN}$

Impact

The toughness of a material is its ability to withstand shock loading. Toughness may be affected by a number of factors including the speed at which a force is applied.

When a load is applied suddenly, instead of gradually, the maximum stress may be increased considerably. This stress produces vibrations which die away, and when equilibrium is reached the stress becomes the ordinary static stress.

An impact test is undertaken on a notched bar. The impact testing machine (see Fig. 1.10) consists of a pendulum, which is raised to an initial starting position where it has potential energy stored due

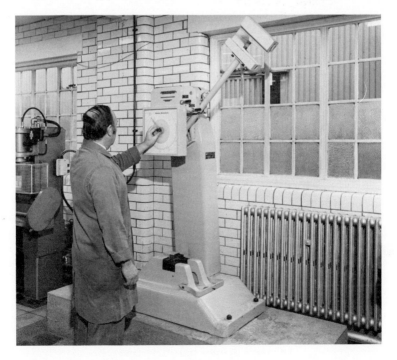

Fig. 1.10 An impact testing machine. *(By courtesy of Avery-Denison Ltd.)*

to its weight and position above the specimen. When it is released, the pendulum swings down, breaks the specimen, and swings through to the other side.

If the specimen is of low toughness, the pendulum will swing through almost to the same height as its starting position. If the specimen is very tough, almost all the energy stored in the pendulum will be absorbed by the specimen and the distance travelled through will not be as great. The amount of energy absorbed is registered on an indicating dial.

Values are given in Izod or Charpy numbers. Some typical Izod values are included in Table 1.3.

Table 1.3   *Typical applications of materials*

Application	Material	Max. tensile stress (MN/m²)	Min. Izod value (Joule)	Brinell hardness (3000 kg 10 mm ball)
Gear wheels, cam shafts, spindles, levers	Case-hardened mild steel EN 32	540	54	153
Loco spindles, armature shafts, lathe spindles	Medium carbon steel (H & T) EN 5	690	54	207
Hard surface with tough core. Better quality gear wheels cam shafts, spindles and levers	Carbon manganese steel (H & T) EN 201	620	108	183
Crankshafts, axles, connecting rods	3% Nickel steel (H & T) EN 21	930	47	277
Pressure vessels, studs and bolts for superheaters working at temperatures up to 550°C	Chromium molybdenum steel (H & T) EN 20	1003	54	293

Hardness

The property of hardness is the ability of a material to resist wear and indentation.

The three principal methods of measuring hardness of materials are:
   (i)   the Rockwell hardness test;
   (ii)   the Brinell hardness test;
   (iii)   the Vickers Pyramid hardness test.

A hardness number is determined by each test which related to the hardness numbers of other known materials. The number increases with the hardness.

The Rockwell test utilises mainly three scales, A, B and C, and the values of hardness are read directly from scales on the machine. Scales A and C utilise a diamond cone loaded with 60 kg and 150 kg respectively. A 1.5 mm diameter hardened steel ball is used with scale B, which is loaded with 100 kg. The penetrator is first loaded with a minor load and the indicator set to zero. The above major load is applied and after its removal the dial gauge records the depth of impression in terms of Rockwell numbers (Fig. 1.11).

(a)   (b)   (c)

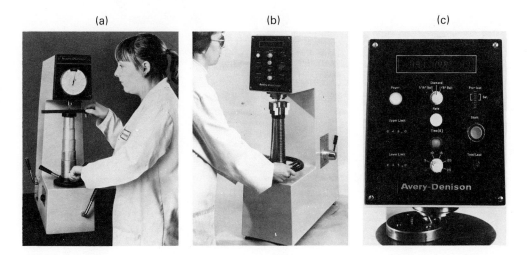

Fig. 1.11   (a) A hardness testing machine to determine Rockwell hardness. *(By courtesy of Avery-Denison Ltd.)*

(b) A digital hardness testing machine to determine Rockwell hardness. *(By courtesy of Avery-Denison Ltd.)*

(c) A control panel for digital testing machine to determine Rockwell hardness. *(By courtesy of Avery-Denison Ltd.)*

The Brinell test consists of pressing a hardened steel ball into the surface of the metal and measuring the diameter of the impression. Then by referring to a table, the hardness number (BHN) is obtained.

These numbers are calculated from

$$BHN = \frac{Load}{Surface\ area\ of\ impression}$$

(see Fig. 1.12).

Fig. 1.12 A Brinell hardness testing machine. *(By courtesy of Avery-Denison Ltd.)*

The Vickers Pyramid test is similar to the Brinell test, the indenter being a pyramid diamond on a square base. The calculation of hardness number is again based on the ratio

$$\frac{\text{Load}}{\text{Area of impression}}$$

The Vickers and Brinell hardness numbers are very similar up to a hardness of 300, but the Vickers test is more reliable than Brinell for harder materials because the diamond does not deform under heavy loading like the steel ball of the Brinell test.

The Rockwell test is particularly useful for rapid routine tests on finished products.

Another method, used to a lesser degree than those above, is the scleroscope. The principle used in the scleroscope is the drop and rebound of a diamond-tipped hammer. The hammer drops by the force of its own gravity from a fixed height on to the test specimen and the resulting rebound is read against the graduated scale. This method is particularly useful where an indentation is undesirable or the surface is inaccessible to the normal hardness testing machines (see Fig. 1.13).

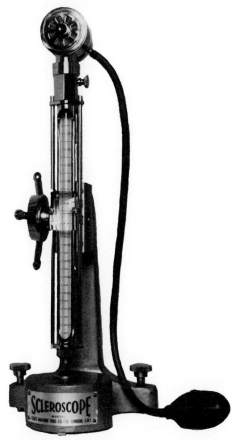

Fig. 1.13 A scleroscope. *(By courtesy of Coats Machine Tool Co.)*

There is direct relationship between hardness and tensile strength, (see Table 1.3). Therefore hardness tests can be used to give an indication of the strength of a metal (see Table 1.4).

Table 1.4 *Comparison of hardness values*

| Material | Rockwell | | Brinell | Vickers | Scleroscope |
	B	C			
Nitrided surface	—	68	745	1050	100
White cast iron	114	44	415	437	57
Soft chisel steel	99	22	235	235	34
Mild steel	74	—	131	131	20
Soft brass	—	—	60	61	—

Fatigue

Many machine parts are subjected to fluctuating stresses. The stress required to cause failure, if it is applied a large number of times, is much less than that necessary to fracture a material with a load which is steadily applied for a limited number of occasions.

In fatigue testing, specimens are subjected to a gradually decreasing range of reversals of stresses, until failure occurs or until ten thousand million cycles have been endured. From the results an *S–N* curve is plotted as shown in Fig. 1.14, where S = stress required for failure and N = number of cycles.

The value of stress at x is that which will cause failure with a gradually applied load and the value y will result in failure if the application of cyclic loading exceeds a million reversals in this instance.

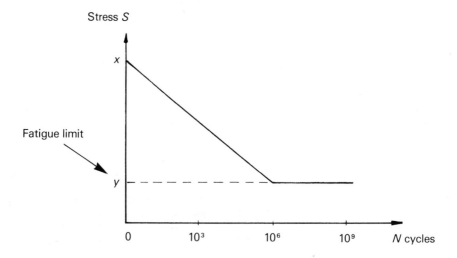

Fig. 1.14   Fatigue testing: $S - N$ curve. This $S - N$ curve is plotted on a logarithmic scale. It should be appreciated that a primary $S - N$ curve is differently shaped. While the curve shown is convenient when determining the fatigue limit, it does not express the true rate of variation of cycles with stress.

THE FACTOR OF SAFETY

We have already seen how stress can be calculated from the size, position and type of loading, the dimensions of the structure and the properties of the materials. In practice, however, none of these factors is known exactly and assumptions are often made.

The dimensions of the members should be known with accuracy although sudden changes of cross-section, which will cause stress concentrations, may arise from an oil-hole, keyway, toolmark or sharp fillet.

The type of load may be static, such as a dust extraction unit mounted on a gantry, or a live load similar to a crane lifting and lowering its load. With a fluctuating load, such as that experienced by a suspension system on a motor car, consideration will have to be given to the fatigue limit of the material in use.

The designer may use his experience and knowledge to estimate the maximum permissible stress which he will accept in his design. Very often he utilises a *factor of safety*, which is a ratio between maximum stress and permissible or working stress.

$$\text{Factor of safety} \ = \ \frac{\text{Maximum stress}}{\text{Working stress}}$$

and hence

$$\text{Working stress} \ = \ \frac{\text{Maximum stress}}{\text{Factor of safety}}$$

The maximum stress of a material determines the maximum load a material can withstand for a given cross-sectional area as:

$$\text{Maximum stress or Tensile strength} \ = \ \frac{\text{Maximum load}}{\text{Original cross-sectional area}}$$

However, a designer would not load a component up to its maximum stress for it would be almost at its breaking point.

The value of the factor of safety can be estimated by the multiple of the following factors:

(i)   the ratio between tensile strength and elastic limit for the material used; generally this value is about 2.

(ii) the type of loading, i.e. static load, 1
      live load, 2
      fluctuating load (giving alternate tensile and compressive stresses), 3

(iii) the system of loading, i.e. gradually applied load, 1
      suddenly applied load, 2
      shock load, 3

(iv) miscellaneous allowance, i.e. chance of unintentional overload,
      use of imperfect material, etc.
      This factor may vary between $1\frac{1}{2}$ and 10.

Example 1 Determine the factor of safety to be used for a steel clamping bolt where the load is static, but could be suddenly applied, and standard specification material is to be used.

Solution   From the factors above,

$$\text{Factor of safety} = 2 \times 1 \times 2 \times 1\tfrac{1}{2} = 6$$

Example 2 Determine the factor of safety for a cast iron wheel rim. The load will be static and gradually applied but the chance of unintentional overload is very high.

Solution

$$\text{Factor of safety} = 2 \times 1 \times 1 \times 10 = 20$$

The factor of safety must never be low enough to allow the working stress to exceed the limit of proportionality stress (see Table 1.5 and Fig. 1.15), for at this point stress is no longer proportional to strain when a further load is applied.

Table 1.5   *Factors of safety for general use*

Material	*Static load*	*Live load with shock*
Cast iron	6	20
Steel	5	12

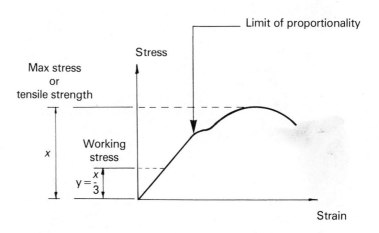

Fig. 1.15   The working stress permitted with a factor of safety of 3.

Example 1 A bolt is manufactured from material which has a maximum tensile strength (stress) of $380\,MN/m^2$. Find the minimum diameter of the bolt which will support a tensile axial force of 42.5 kN with a factor of safety of 6.

Solution   Working stress $= \dfrac{\text{Tensile strength}}{\text{Factor of safety}} = \dfrac{380\,MN/m^2}{6} = 63.33\,MN/m^2$

N.B. $63.33\,MN/m^2 = 63.33\,N/mm^2$

Cross-sectional area $= \dfrac{\text{Force}}{\text{Stress}} = \dfrac{42\,500\,N}{63.33\,N/mm^2} = 671.1\,mm^2$

$$\text{Area} = \frac{\pi d^2}{4}$$

$\therefore$ Diameter $= \sqrt{\dfrac{4 \times \text{area}}{\pi}} = \sqrt{\dfrac{4 \times 671.1}{\pi}} = \sqrt{854.3}$

$$= 29.23\,mm$$

The minimum diameter of the bolt is (in practice) 30 mm.

Example 2 A wire rope, 25 mm diameter, is manufactured from material which has a maximum tensile strength of $500\,MN/m^2$ and is used to support a lift. The lift has a mass of 500 kg and is to carry a maximum of 15 persons, of average mass 60 kg each. Calculate (i) the factor of safety used in this design and (ii) the additional force required to accelerate the lift at a constant rate of $1.5\,m/s^2$.

Solution   Total mass to be lifted $=$ Mass of lift $+$ Mass of people

$$= 500\,kg + (15 \times 60)\,kg$$

$$= 1400\,kg$$

Total force exerted by this mass $= 1400\,kg \times 9.81\,m/s^2$

$$= 13\,734\,N$$

(i)   Working stress $= \dfrac{\text{Force}}{\text{Cross-sectional area}} = \dfrac{13734}{\dfrac{\pi \times 25^2}{4}}$

$$= 27.98\,N/mm^2$$

$\therefore$ Working stress $= 27.98\,MN/m^2$

Factor of safety $= \dfrac{\text{Tensile strength}}{\text{Working stress}} = \dfrac{500}{27.98} = 17.87,$

say 18

(ii) $$\text{Accelerating force} = \text{Mass} \times \text{Acceleration}$$
$$= 1400 \, \text{kg} \times 1.5 \, \text{m/s}^2$$
$$= 2100 \, \text{N}$$

Example 3 Find the maximum axial compressive load which can be supported by a cast iron tube, 200 mm outside diameter and 160 mm inside diameter. The maximum compressive stress for cast iron is 535 MN/m² and a factor of safety of 3 is to be used.

Solution $$\text{Cross-sectional area supporting load} = \frac{\pi}{4}(200^2 - 160^2)$$
$$= 11\,311 \, \text{mm}^2$$

$$\text{Working stress} = \frac{\text{Tensile strength}}{\text{Factor of safety}}$$

$$= \frac{535}{3} = 178.33 \, \text{MN/m}^2 = 178.33 \, \text{N/mm}^2$$

$$\text{Maximum compressive load} = \text{Stress} \times \text{Area}$$
$$= 178.33 \, \text{N/mm}^2 \times 11\,311 \, \text{mm}^2$$

$$\therefore \text{Maximum compressive load} = 2.02 \, \text{MN}$$

CORROSION

The Cost of Corrosion

An important aspect of design is the consideration of whether a metal will corrode.

A report published by a government committee in 1971 estimated that corrosion was costing the country £1365 million a year, and in the committee's opinion at least a quarter of this sum could be saved by the use of well-established corrosion-control techniques.

Consider the following results that may occur from uncontrolled corrosion:

   (i) continuous damage to process plant, structural assemblies and other equipment;
  (ii) consequent shutdowns for repair or replacement work;
 (iii) the risk of injury to personnel;
  (iv) contamination of the product;
   (v) the loss of the product;
  (vi) the loss of operating efficiency;

(vii) unfavourable publicity associated with hazards to the public and/or the environment;

(viii) customer alienation.

Bearing the above points in mind the designer must consider carefully the most suitable material and corrosion-control methods for a given application involving a corrosive environment. He or she should not lose sight of the fact that corrosion control is only one of several factors which have to be assessed in material selection.

Other factors such as the cost of material, processing costs, mechanical and physical properties may also have to be considered.

The following paragraphs will help the reader understand the processes of corrosion and the preventative techniques that can be used against it.

What Corrosion Is

Corrosion is the gradual eating away of the metal and the speed at which a metal corrodes will depend upon:

(i) The environment in which the metal part is operating.
For example, iron will rust much quicker if suspended in salt water than if it is suspended in air-free distilled water.
If iron is kept in dry air it will not rust as quickly as when the air is moist.

Water contaminated by factory waste may accelerate the corrosion of metallic parts considerably. The atmosphere itself may be polluted by harmful corrosive agents. The fumes given off from factory chimneys can in themselves cause corrosion.

(ii) The nature of the metal itself.

Some metals when in contact with the atmosphere react chemically with it so that an impervious homogeneous coating forms on the metal's surface protecting it from further attack by the atmosphere. Aluminium, copper, lead, magnesium and zinc are such metals.

Unfortunately the oxide coating that forms on iron and plain carbon steel is porous and does not protect the metal from further attack by the atmosphere. Therefore corrosion is progressive.

Corrosion Resistance Using Alloys

Metals may be alloyed to improve their corrosion resistance. Consider the following corrosion-resistant alloys.

Stainless Steels

There are three main types: the martensitic and ferritic grades containing 11–18% chromium, and the austenitic grades containing approximately 17–26% chromium and 8–22% nickel. The highest general corrosion resistance is obtained with the austenitic grades.

Stainless steels are resistant to atmospheric corrosion, nitric acid, some concentrations of sulphuric acid, many organic acids and, under certain conditions, sulphurous acid and alkalis.

Copper Alloys

Copper resists sea water, hot or cold fresh water, non-oxidising acids and atmospheric attack. Certain alloying additions improve its mechanical and physical properties and also its corrosion resistance.

Hence the use, for example, of aluminium brass and cupronickels as condenser-tube alloys, and of the aluminium bronzes for such applications as pump bodies, pickling cradles and ships' propellers.

Aluminium Alloys

Aluminium offers good resistance to atmospheric corrosion and to many other corrosive media (e.g. acetic acid, nitric acid, fatty acids, sulphur and sulphurous atmospheres). It is alloyed with small amounts of other metals mainly to obtain improved mechanical and physical properties.

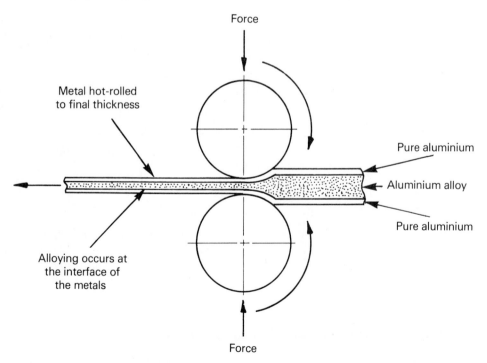

Fig. 1.16 Cladding.

The aluminium–magnesium and aluminium–manganese alloys are generally regarded as exhibiting the highest corrosion resistance, followed by the aluminium–magnesium–silicon and aluminium–silicon alloys.

The aluminium–copper alloys are the least resistant to corrosion. Their corrosion resistance can be improved however by cladding the alloy with a thin layer of pure aluminium (Fig. 1.16).

Nickel Alloys

Nickel is resistant to hot or cold alkalis, dilute non-oxidising inorganic and organic acids and to the atmosphere. It is widely used in electroplating as a corrosion-resistant layer which is then covered with a thin film of chromium.

The addition of molybdenum and occasionally chromium improves its resistance to hydrochloric, sulphuric and phosphoric acids, chlorine gas and other acid and oxidising environments.

Titanium Alloys

Titanium and its alloys are extremely resistant to corrosion, particularly in sea water and in industrial atmospheres. These alloys are often used in chemical plant.

The Process of Corrosion

The corrosion process may be likened to the electrolytic action that takes place in a simple cell. When two dissimilar metals are in contact in the presence of an electrolyte, one having a higher electrode potential than the other, an electric current will flow between the metals.

Metals placed in order of their electrode potential, may be called the electrochemical series (Table 1.6).

A simple corrosion cell (Fig. 1.17) will consist of the following:

(i)   the *anode,* the site at which the metal is corroded;

(ii) the *cathode,* which is not consumed in the corrosion process;

(iii) an electrically conducting solution called the *electrolyte,* which is the corrosive medium.

The corroding anodic metal passes into the electrolyte as positively charged ions, releasing electrons which participate in a reaction at the cathode. Therefore the corrosion current between the anode and cathode consists of electrons flowing within the metal and ions flowing within the electrolyte.

The reader may confirm the formation of current by setting up a simple corrosion cell similar to the one in Fig. 1.17 using zinc and copper plates and an electrolyte of dilute sulphuric acid.

Table 1.6   *The electrochemical series*

	Metal	Electrode potential (V)	
Base	Magnesium	−2.38	Anodic
	Aluminium	−1.67	(tending to corrode)
	Zinc	−0.76	
	Chromium	−0.56	
	Iron	−0.44	
	Cadmium	−0.40	
	Nickel	−0.23	
	Tin	−0.14	
	Lead	−0.12	
	Hydrogen	0.00	
	Copper	+0.34	
	Silver	+0.80	Cathodic
Noble	Gold	+1.36	(more protected)

Note. The potential difference between a metal electrode and a hydrogen electrode is called the electrode potential.

The zinc, being the more anodic of the two (see Table 1.6), will visibly corrode. An electric current will flow between the zinc anode and the copper cathode and may be measured using a milliammeter.

The rate of corrosion is influenced by the electrical conductivity of the electrolyte. The corrosion rate being high in good conductors, e.g. salt water and acid solutions, and low in poor conductors, e.g. high-purity water.

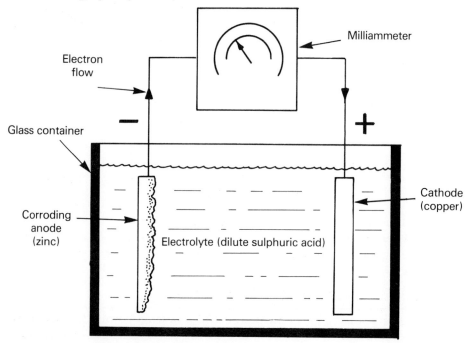

Fig. 1.17   A simple corrosion cell.

This again may be confirmed by the reader, if he or she substitutes the dilute sulphuric acid in the simple cell with distilled water, and notes the difference in the corrosion rate of the zinc and the current flowing between the two plates.

The whole surface of a metal may be anodic or cathodic as would be the case if two dissimilar metals were placed together. It should be noted that if these two metals are widely spaced in the electro-chemical series (see Table 1.6) and an electrolyte is present, the greater will be the corrosion current and the greater the corrosion of the more anodic metal.

In the case of surface corrosion, very small electric currents are set up between localised anodic and cathodic areas that occur in the same metal surface due to, for example, slight differences in composition (Fig. 1.18).

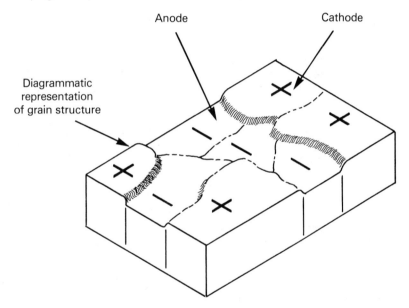

Fig. 1.18   An impression of local anodes and cathodes of corrosion cells occurring on the same metal surface. *(By courtesy of the Department of Industry, Committee on Corrosion.)*

A component may be immersed in a liquid solution during service. This liquid may act as an electrolyte, or the electrolyte may be a thin film of moisture formed perhaps by rainwater or condensation.

Methods of Controlling Corrosion

In the war against corrosion the designer and engineer have at their disposal the following preventative techniques, all of which attempt to interfere with the corrosion mechanism.

Modifying the Environment

As we have seen earlier in the chapter the rate of corrosion may be related to the electrical conductivity of the electrolyte.

i.e.   the greater the electrical conductivity of the electrolyte

↓

the greater the current flow

↓

the greater the rate of corrosion

By reducing the conductivity, and hence the current flow within the electrolyte, the rate of corrosion may be slowed down.

The most common factor in the corrosion process is the presence of water in the environment. This acts as an electrolyte.

The conductivity of water used for industrial and other processes may be reduced by:

(i) The removal of dissolved gases (particularly oxygen, which contributes to the rapid corrosion of iron and steel). This can be achieved by chemical de-activation or de-aeration, which involves holding the water at low pressures, raising its temperature, or purging it with an inert gas.

(ii) The use of chemical inhibitors which stifle the anodic and cathodic reactions.

Moisture and contaminants in the atmosphere which cause corrosion are very difficult to control. Indoors the atmosphere can be filtered, cleaned and dried by an air-conditioning system.

Metal may corrode if it is buried in the soil, due to moisture and dissolved salts, and control of this environment is again difficult.

Protection against corrosion due to the atmosphere and soil conditions is best achieved by one of the techniques explained in the following paragraphs.

Controlling Corrosion by Electrical Methods

(i) *Cathodic Protection.* Corrosion may be controlled by making the metal, which is to be protected, more cathodic with respect to an external anode.

This is achieved by supplying an external low-voltage direct current.

The positive terminal is connected to an auxiliary anode (e.g. scrap iron) located some way from the structure to be protected and the negative terminal to the structure itself.

Steel, copper, lead and brass are among the metals which may be protected in this way.

Applications include pipelines, tanks, and bridge and pylon footings (see Fig. 1.19).

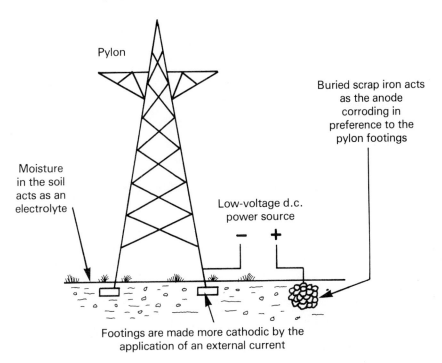

Pylon

Buried scrap iron acts as the anode corroding in preference to the pylon footings

Moisture in the soil acts as an electrolyte

Low-voltage d.c. power source

Footings are made more cathodic by the application of an external current

Fig. 1.19 The cathodic protection of pylon footings.

(ii) *Sacrificial Cathodic Protection.* A structure can also be protected against corrosion by a sacrificial anode. If an auxiliary electrode is made of a metal which is more anodic than the metal to be protected, it will become the anode of a simple corrosion cell and a current will flow between it and the structure to be protected, i.e. the cathode; the structure will be protected sacrificially by the corroding anode.

The sacrificial anode may be made from magnesium, magnesium-base alloys, zinc and aluminium.

Because the corrosion cell creates its own electric current, this method of corrosion protection is of particular value where an external source of current is inconvenient. Protection of course ceases if the anode corrodes away completely.

An application of this principle is used on a ship's hull (see Fig. 1.20).

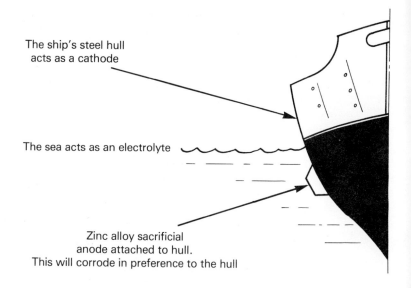

The ship's steel hull
acts as a cathode

The sea acts as an electrolyte

Zinc alloy sacrificial
anode attached to hull.
This will corrode in preference to the hull

Fig. 1.20   The sacrificial cathodic protection of a ship's hull.

Controlling Corrosion by the Use of Protective Coatings

Coatings are used to protect metal from corrosion in one or more of the following ways.

(i) *Exclusion of the Environment*. All coatings exclude the environment from contact with the metal to a certain extent, but to be effective those that protect by exclusion alone must cover the surface completely and must be sound and resistant to mechanical damage. Straightforward excluders are vitreous enamels, lacquers, non-inhibited paints (paints that do not have corrosion inhibitive qualities), plastics and metal coatings (such as nickel and chromium on steel) which are *more* noble than the metal to which they are applied. Soundness is essential in metallic excluder coatings, otherwise the (cathodic) coating will promote corrosion of the (anodic) underlying metal at any discontinuities (Fig. 1.21).

(ii) *Sacrificial Coatings*. These act in two ways: where the coating is sound they act as excluders; and where discontinuities occur they provide cathodic protection, just as a sacrificial anode protects a buried or immersed structure. In this case the coating is *less* noble under the prevailing corrosive conditions than the metal being protected, and therefore the anodic coating would corrode in preference to the underlying metal.

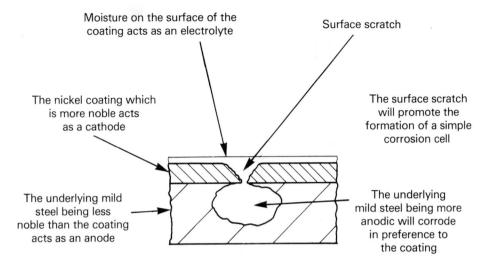

Fig. 1.21 Imperfect nickel coating on mild steel.

An example of this would be the sacrificial protection given to mild steel sheet by a coating of zinc (galvanised sheet), as in Fig. 1.22.

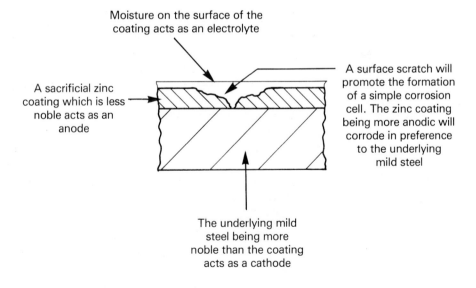

Fig. 1.22 Sacrificial zinc coating on mild steel.

Sacrificial coatings must be thick enough to last the service life of the component, for they corrode slowly themselves and are consumed as they protect.

(iii) *Inhibited Coatings.* These organic coatings are primarily priming paints, oils and greases which contain a corrosion inhibitor in suspension (as part of the pigment) or in solution (oil or grease).

They act primarily as excluders, but since no organic coating is totally impervious to moisture, they also provide a reserve of corrosion inhibitor at the metal surface to protect against moisture penetration.

Some of the more important methods and types of protective coating will now be considered.

(i) *Metallic Coatings.* These coatings may be further subdivided.

(a) Metal Spraying. This technique is used widely to apply coatings (e.g. zinc, aluminium and their alloys) to steel and aluminium.

The metal, in the form of wire or powder, is fed through a flame, atomised by a jet of air, and then projected on to the prepared surface.

(b) Cladding. This is applicable primarily to sheet metal. In this case the metal to be protected is sandwiched between layers of the coating metal and then hot-rolled to final thickness. Alloying occurs at the interface of the metals during rolling (Fig. 1.16).

Cladding is often employed either to provide sacrificial protection (e.g. certain types of aluminium alloy clad with aluminium) or to economise on a costly corrosion-resisting alloy by using it as a cladding on a cheaper material which supplies structural reinforcement (e.g. carbon steel clad with stainless steel).

(c) Hot Dipping. The cleaned component is immersed in a molten bath of the coating metal (usually zinc, tin or aluminium), resulting in a coating with very good adhesion and coverage (even in crevices).

(d) Cementation. The outer surface of the part to be protected is converted into an alloy of higher corrosion resistance by being heated in contact with zinc powder (sheradising), aluminium powder (calorising) or a gaseous compound of chromium (chromising). Layers of considerable thickness can be achieved.

(e) Electroplating. Many metals can be plated on to a variety of other metals by electroplating, also called electrodeposition.

Electroplating is chosen as a method of protection when the coatings are to be relatively thin and of controlled thickness, or when coating processes involving high temperatures have to be avoided. Typical electrodeposits are nickel, tin, zinc, cadmium, copper, silver and chromium. Chromium plating for example is used to protect and to give a pleasing finish to steel and zinc-base components on motor cars. Unfortunately the very thin chromium deposit may contain very small cracks and in practice the component is first plated with nickel and then with chromium, so that during service the nickel protects the underlying metal at cracks in the chromium.

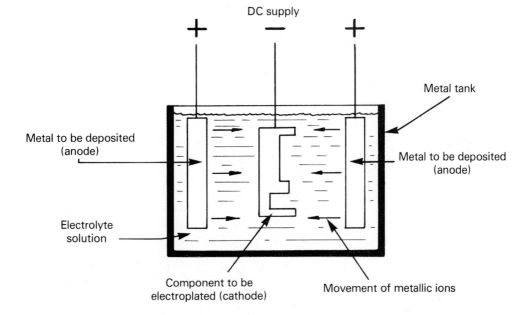

Fig. 1.23 The principle of electroplating.

The principle of electroplating is shown in Fig. 1.23. The component to be electroplated is made the cathode, this being connected to the negative terminal of a d.c. supply. The plating material being the anode is connected to the positive terminal and both are immersed in an electrolytic solution.

Metallic ions are removed from the anode into the electrolyte and deposited from the electrolyte on to the cathode.

Note. In most cases the plating material is more noble (naturally more cathodic) than the underlying metal which would be more base (naturally more anodic).

By using a stronger e.m.f. than the natural e.m.f. set up between two dissimilar metals in the presence of an electrolyte, in the case of the simple corrosion cell (Fig. 1.17) the natural direction of the electron flow may be reversed, making the naturally cathodic material more anodic.

(ii) *Inorganic Coatings.* All inorganic coatings are brittle and therefore involve risk of corrosion at cracks, defects and weaknesses in the coating. It is virtually impossible to avoid imperfections, but by superimposing cathodic protection, corrosion at defects can be prevented.

(a) Vitreous Enamels. Vitreous enamel coatings are obtained by applying a glass powder to the surface of the component and heating so that the glass fuses to the metal. These coatings are used mostly on steel and cast iron. They offer good resistance to the atmosphere and various other environments, and are employed in such applications as decorative finishes (e.g. on street signs and architectural panels), tank coatings, baths and other domestic equipment, etc.

(b) Cement. Cement coatings are cheap, have a coefficient of expansion approximating to that of steel, and can be easily applied by centrifugal means, spraying, or with a trowel, and are readily repaired. Thick coatings generally incorporate a mesh reinforcement. Cement coatings can be used to protect steel or iron water pipes and the interior of hot and cold water tanks and chemical storage tanks.

(iii) *Conversion Coatings.* These are produced by treating the metal surface chemically with an appropriate solution. There are three main types.

(a) Phosphate Coatings. Phosphating is applied principally to steel, but also to zinc and cadmium. Steel is treated in a weak phosphoric acid solution of iron, zinc or manganese phosphate.

Phosphate coatings give only limited protection, but provide an excellent base for paint or other protective coatings.

(b) Chromate Coatings. These coatings can be produced on aluminium and its alloys, and cadmium and zinc. They offer a useful degree of resistance to corrosion and a good preparation for painting.

(c) Anodised Coatings. The thin invisible oxide coating that forms naturally on zinc, aluminium and their alloys gives them some protection against corrosion. In anodising this oxide coating is increased considerably.

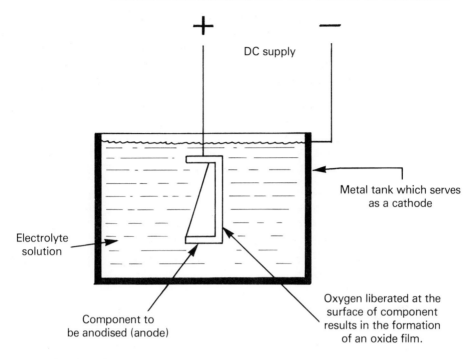

Fig. 1.24   The principle of anodising.

Anodising (Fig. 1.24) is an electrolytic process in which the component is made the anode in a circuit (not the cathode as in electroplating), and suspended in a suitable electrolyte of chromic or sulphuric acid.

The tank in which the process is carried out serves as the cathode.

The chemical reaction which takes place at the anode liberates oxygen. This results in the formation of an oxide film of considerable thickness.

The porous nature of this film enables it to be dyed a variety of colours.

Anodising is also a good pre-treatment for painting.

(iv) *Organic Coatings.* These coatings are probably the most familiar means of controlling corrosion of metals. Paints, pitch, tar and bitumen are the most common, but plastics are now also beginning to be widely used.

(a) Paints. In general, paints are used as protection against atmospheric corrosion and not for the protection of metal structures buried in the earth.

Paint consists of an intimate mixture of pigment and liquid medium which, after application, dries (or reacts) to form a coherent coating.

Three coats are generally applied: a *priming coat,* formulated to achieve maximum adhesion, and often containing a pigment, which is a substance with corrosion inhibiting properties (e.g. red lead); an *undercoat,* used essentially to build up the final thickness of the coating system; and a *finishing coat,* designed to provide maximum weathering resistance and the final colour and texture.

In recent years the use of high-build-up paint systems has grown considerably. These coatings are thick enough to act as a very substantial long-term barrier between the metal being protected and its environment.

(b) Plastic Coatings. Plastics provide good protection against corrosion while at the same time enhancing the appearance of the component.

Plastic coatings can replace painting, plating, anodising and galvanising.

Typical plastics used as coatings are epoxy resins, polyurethanes, PVC, nylon and polyethylene.

Most plastic coatings are applied by immersing prepared and heated metal into fluidised coating powders or liquid materials, although certain plastics may be applied by brush or spray.

The full process involves degreasing, shot blasting, application of a suitable primer, heating, dipping and curing.

Surface Preparation

A coating protects only as well as it adheres. Whatever the coating system, it is essential to remember that cleaning, and surface preparation of the component prior to coating, is as vital as the coating process itself.

Typical cleaning processes are as follows.

(a) Blast Cleaning. This is a process in which an air or water-borne shot is fired at the surface of the component. Dirt, oxide film and scale can be quickly removed by this process. Common blast-cleaning materials are: iron grit; iron shot; steel shot; crushed fused bauxite; and glass beads.

(b) Electrolytic Cleaning. This process is used to remove oil, grease or dirt from the surface of a component. The component to be cleaned is immersed in a suitable cleaning solution contained in a tank. The component is made the cathode and the tank the anode. A low-voltage electric current is put through the circuit and large volumes of hydrogen are given off from the cathode removing any dirt or grease that may be present.

(c) Pickling. Oxide scale and rust may be removed by immersing the components in a heated pickling solution of sulphuric or hydrochloric acid.

Inhibitors are often added to the pickling solution to minimise chemical corrosion of the component. The component must be washed after pickling.

(d) Flame Cleaning. Old paint, oil, grease and mill scale may be removed easily using a neutral oxyacetylene flame.

(e) Abrasive Cleaning. A surface may be prepared for coating treatments using abrasive cleaning methods. These include the use of grinding wheels and abrasive bands.

Having considered the choice of materials and corrosion protection technique to fulfil an intended purpose, the designer should not forget that corrosion may be minimised by careful design of the component. For example, components should not contain pockets or crevices in which water or other corrosive liquids could be trapped. Consideration must be given to dissimilar metals in contact with each other. Drain holes and drain cocks should be provided where necessary. Access to parts liable to corrode should be 'designed in' to permit adequate initial protection and also maintenance during service.

Some typical design faults and their remedies are given in Fig. 1.25.

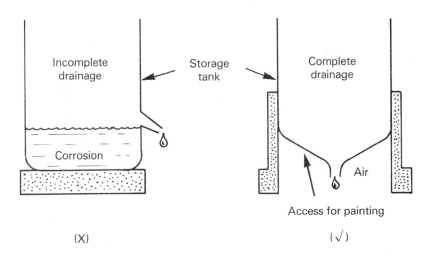

Fig. 1.25 Design faults and their remedies: (√) = good design; (X) = bad design *(continued overleaf). (By courtesy of the Department of Industry, Committee on Corrosion.)*

Collection of water in a structural member

(X)

Provide drainage holes or reverse the member.

(√)   (√)

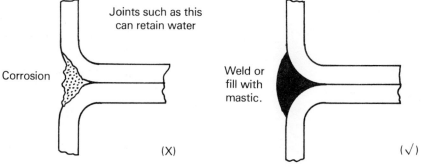

Overlap joint

Water

Corrosion

(X)

Water

Water cannot
enter the
joint

(√)

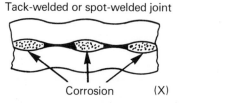

Joints such as this
can retain water

Corrosion

(X)

Weld or
fill with
mastic.

(√)

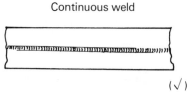

Tack-welded or spot-welded joint

Corrosion   (X)

Continuous weld

(√)

Fig. 1.25 *(continued)*

PROPERTIES OF MATERIALS NECESSARY FOR MANUFACTURING PROCESSES

Materials may be classified as (i) wrought materials or (ii) casting materials.

A wrought material is a material that is capable of being worked to shape in the solid state, either hot or cold, by such processes as rolling, drawing, forging and extrusion.

A casting material is one which is heated to a molten state and poured or forced into a suitably prepared mould, typical casting processes being sand-casting, die-casting and injection moulding. These processes are called *primary forming processes,* and some of them will be considered in more detail in Chapter 2.

Materials may be joined by various thermal techniques such as soldering and welding.

When selecting a primary forming process or other process such as bending or welding, the designer must consider the mechanical and physical properties of the material, as all materials cannot be formed in the same way.

There are a number of typical properties that should be considered, which we shall now discuss.

Plasticity

This is the ability of a material to be deformed by hot or cold working.

The material is caused to 'flow' under the action of a force, in much the same way as a viscous liquid would flow.

Metal undergoes plastic deformation in the following operations: drawing, bending, rolling, extrusion and forging (see Fig. 1.26).

Ductility

This is the ability of a material to undergo cold plastic deformation by drawing out in tension.

This is a desired property of materials that are to be formed by drawing operations. The cross-sectional area of the material is reduced and its length increased by:

(i) pulling through a metal die (Fig. 1.26(a)), or

(ii) pressing into shape using a press tool. Fig. 1.26(b) shows a deep drawing operation.

Materials that are to be bent to shape also need to be ductile (Fig. 1.26(c)).

(a) Drawing

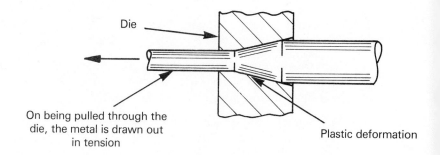

Die

On being pulled through the
die, the metal is drawn out
in tension

Plastic deformation

(b) Deep drawing

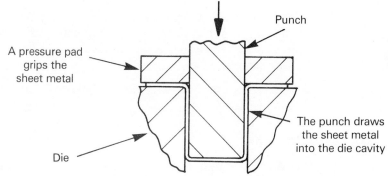

Punch

A pressure pad
grips the
sheet metal

The punch draws
the sheet metal
into the die cavity

Die

(c) Bending

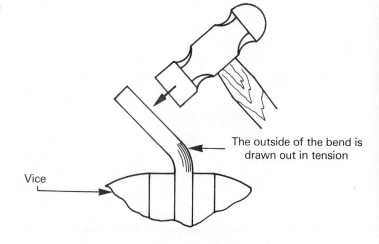

The outside of the bend is
drawn out in tension

Vice

Fig. 1.26   The plastic deformation of materials.

(d) Rolling

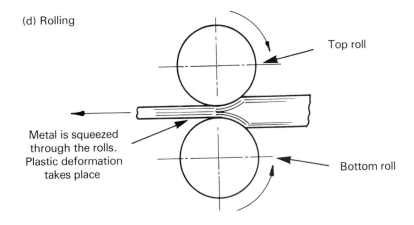

Top roll

Metal is squeezed
through the rolls.
Plastic deformation
takes place

Bottom roll

(e) Extrusion

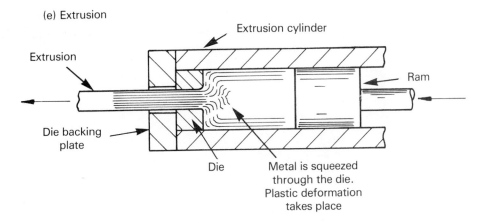

Extrusion cylinder

Extrusion

Ram

Die backing
plate

Die

Metal is squeezed
through the die.
Plastic deformation
takes place

(f) Drop forging

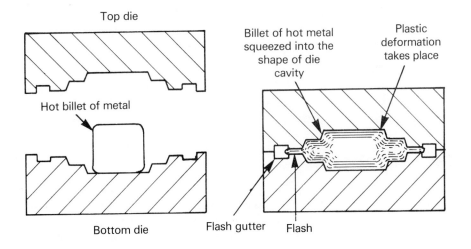

Top die

Hot billet of metal

Billet of hot metal
squeezed into the
shape of die
cavity

Plastic
deformation
takes place

Bottom die

Flash gutter

Flash

Fig. 1.26 *(continued)*

Malleability

This is the ability of a material to undergo cold plastic deformation by hammering, rolling or squeezing. The cross-sectional area of the metal is reduced and its length or width or both are increased. Metals that are to be shaped by rolling (Fig. 1.26(d)) and extrusion (Fig. 1.26(e)) require this property.

Brittleness

A material that does not undergo visible plastic deformation and fractures easily is said to be 'brittle'. Another name for brittleness is 'shortness'.

A material that is brittle at high temperatures is said to be 'hot short'.

A typical metal that is 'hot short' is cast iron. Any attempt to forge this metal at a high temperature would result in the metal fracturing.

Brittleness at low temperatures is sometimes called 'cold shortness'.

Fusibility

This is the ability of material to melt easily. Metals that are to be cast or fusion-welded must exhibit this property.

Fluidity

If an intricate casting is required, then the molten metal must be very fluid and run easily. This will allow it to reach every crevice in the mould cavity.

The relatively large amount of carbon in cast iron, approximately $3\frac{1}{2}\%$, helps this metal to become very fluid in the molten state thus allowing intricate castings to be produced.

Plain carbon steel on the other hand contains much less carbon, approximately 0.1–1.5%. The steel therefore does not become very fluid in the molten state, and intricate plain carbon steel castings cannot be produced.

Machinability

At some time during the manufacture of an article certain of its components may have to be machined. This usually occurs after some primary forming process. If the material from which the component is made can be machined with ease, and at the same time a good finish can be produced, then the material is said to have good machinability.

The machinability of a material may depend upon such factors as the type of cutting tool material, the cutting tool geometry, depth of cut, feed, the use of a cutting fluid and the properties of the material itself.

The machinability of a soft ductile metal may be quite poor, in that localised tearing instead of cutting takes place, resulting in a poor finish. If the metal is work-hardened or alloyed with another metal or material, to increase the hardness slightly, the machinability will be improved. The metal should not be too hard, or the cutting tool will not penetrate it, and machining then becomes difficult. For ideal cutting conditions the material should have a low hardness together with low ductility.

Lead is often added to low carbon steel and brasses to improve their machinability. These materials are then called 'free-cutting' materials.

COST PROPERTIES

The profit made by a company is the difference between the total revenue from the sales of the company's products and the total costs incurred in producing them,

i.e. $$£\,Profit\ =\ £\,Sales - £\,Costs$$

The selling price of a product may be determined by

(i) comparing a competitor's selling price for a similar product. For example, if a foreign motor manufacturer decides to introduce a new car into this country, which is to be in direct competition with the Mini Metro, then it needs to have a similar selling price or possess such additional features as would clearly justify a higher price, e.g. better performance or greater economy.

(ii) deciding how much the market is prepared to pay for such a product. For example, if British Telecom decided to introduce a new household telephone, which would operate in both sound and vision, a selling price would need to be determined which would encourage people to buy it. At the same time British Telecom would expect to make a profit from its production and sale.

When selling prices are fixed each product must bear a fair share of the costs/expenses involved in production. If the price is fixed too low, the costs of production will not be recovered and the company may be unable to meet its bills. If the price is too high, the product will be uncompetitive and the market will be reduced. It is important to involve both costing and marketing departments when fixing the selling price to ensure that these requirements are met.

Cost properties will include the price of:

Materials. While silver and copper are better electrical conductors than aluminium, their costs may be prohibitive in the manufacture of a product. It may be more prudent to use aluminium.

Labour. The amount the company pays to its employees who manufacture the product.

Tooling. Large quantity production requires the manufacture of tooling.

Packaging. The product may need to be sold in a plastic or cardboard container.

Servicing. If the cost of servicing is too high, it may limit the future sales of a product.

Distribution and Warehousing. The costs involved in moving the product from its place of manufacture, its storage and delivery to point of sale.

Advertising. 'It pays to advertise.' Unless prospective buyers know of the availability of your product, sales will be limited.

Agencies. The sale of a company's products may be entrusted to another selling organisation. This will increase the costs to the company but it would expect to increase sales.

Let us now look at some of these costs in more detail.

Material Costs

There are two elements of material costs:

(i) *Direct material costs:* those costs for materials which are directly part of the component being manufactured; and

(ii) *Indirect material costs:* materials used in a firm which are not part of the component being produced — oils, greases, tools, wipers — but are a necessary requirement of a production unit.

Direct Material Costs

When the designer makes a decision about the type of material from which an article is to be made, one important factor which he has to consider is the purchase cost of the material. In some manufacturing industries the material cost may account for as much as 50% of the total manufacturing costs of the finished product.

The designer should be aware of current material prices and the various forms of supply in which materials are available — cold-drawn, hot-drawn, cold-rolled, hot-rolled, cast, extruded section, etc.

The form of supply will influence the purchase price of the material — cold-rolled bright stock is more expensive than hot-rolled black stock. The designer will consider the mechanical properties and dimensional accuracy of the material. The use of bright instead of black stock may allow a machining operation to be omitted, thus offsetting the additional material cost, giving a saving in machining time and perhaps thus giving an overall saving in manufacturing costs. A comparison of material costs is given in Table 1.7.

Table 1.7   *Comparison of material prices*

Material	£ cost/kg	£ cost/m³
Metals		
Bright drawn mild steel	0.65	5 070
Black mild steel	0.45	3 510
Bright drawn EN24T steel	1.50	11 700
Stainless steel EN58J	2.40	18 720
Aluminium	3.90	30 420
Brass	2.86	22 308
Copper	3.91	30 498
Phosphor bronze	5.84	45 552
Plastics		
Nylon	7.50	8 253
PVC	3.64	5 098

Note: It may be seen from the above table that the cost/kg of plastic materials is much higher than that for metals. The density of plastics, however, is less than that of most metals, therefore the cost/m³ of plastics is considerably lower in some cases than that of metals. (Prices applicable at Nov. 1987.)

Labour Costs

This is the amount the company has to pay to its employees for producing the goods. Labour costs are usually split into *two* elements.

   (i) *Direct labour costs:* the wages paid to employees who are actually producing the goods on the shop floor, i.e., machinists, fettlers, production fitters, painters, etc.

   (ii) *Indirect labour costs:* the wages and salaries paid to employees who are not producing but are necessary in order that the company can operate successfully, i.e., foremen, wages clerks, salesmen, managers. These costs are generally included in the factory 'overhead costs'.

Example   A machinist machines 150 identical components in a 40 hour week. If his weekly wage is £120 per week, what is the direct labour cost per component?

Solution   Direct labour cost per component $= \dfrac{£120}{150 \text{ components}} = 80\,\text{p}$

Tooling Costs

If a component is to be machined from bar or tube, no tooling costs will be involved other than nominal costs for the cutting tools. However, a great deal of scrap may be produced, by way of turnings or millings, etc, which will involve the company in further costs: (a) in material wastage, and (b) in extra machining costs (increased labour costs). If a small batch of components is to be produced, this method of manufacture may be the most economic. However, if large quantities are to be produced, it will be more advisable to investigate alternative pre-form methods of production, i.e., by casting, pressing, forging, extruding, welding or plastic moulding the component.

Tools (dies or moulds) would have to be produced in order that one of these methods could be adopted. This would involve the company in extra costs to produce the necessary equipment, *tooling costs*, and would be a direct expense incurred in producing the component.

Example   Suppose, for instance, we are to produce a component by the pressure die-casting process (this process is examined in Chapter 2). The cost of the necessary dies for production is £7500. The estimated maximum life of the dies is 250 000 components (when 250 000 components have been produced, new dies will be needed).

Solution   The additional cost, which must be added to each component, to cover the cost of dies $= \dfrac{£7500}{250\,000} = 3\,\text{p}$

Particular advantages will accrue from the manufacture of tooling, and the consequent volume production–labour costs will be reduced as the product/part will be made quicker by less-skilled operatives. However, with this type of production there is a need for inter-changeability — all parts produced by one machine must be easily assembled with parts produced by another machine — or the advantage of volume production may diminish.

Packaging Costs

Packaging has an important part to play in the merchandising of the product and is classified as a direct expense.

The packaging of say a range of drills has to be substantial enough to make sure they are contained during transit and sale.

For consumer and luxury articles the packaging has to be designed to help 'sell' the product. Perfume is sold in packaging which is expensive to produce, but the cost of it is justified in the selling price of the product.

Servicing Costs

During service the manufactured article will need to be maintained in order that it can continue to perform its function satisfactorily.

The cost of this maintenance is a *servicing cost*.

The cost of servicing has to be borne in mind when a product is designed. It may well be that servicing of any kind may prove too expensive, and it may be more economic to use the product until it is faulty and then discard it. This practice does occur with some low-cost items such as printed circuit boards for electronic equipment, throw-away tools and disposable consumer items.

Direct and Overhead Expenses

We have seen that when manufacturing a product, as well as the cost of labour and materials, we also have to incur other expenses, such as the cost of tooling, packaging and servicing, and these may be included in *direct* or *overhead expenses*.

Direct Expenses

These are expenses which are specially attributable to production costs, other than direct labour and direct material costs. They may include the costs of tooling, packaging or subcontracting the work to another firm.

Overhead Expenses

All indirect costs incurred in connection with the production, distribution, sale and administration of the company's products and business are classified as overhead expenses.

These costs will include indirect labour, indirect materials, telephone charges, stationery, general rates, heating, lighting, warehouse costs, transport costs, advertising and promotion, etc.

Example   A product is to be manufactured and a selling price has been agreed. The total expenses of production are made up of *fixed costs* — the costs which have to be paid before production commences, i.e. tooling costs, packaging costs and an appropriate share of overhead expenses — and *variable costs* — the costs of producing each component, i.e. direct labour and direct material costs.

Selling price   £1.50 each

Fixed costs   £12 500

Variable costs   £0.75 each

Draw a break-even chart and determine the quantity of sales needed before the company starts to get a profit on its investment.

Solution   The revenue graph will be a straight line.

To obtain two points determine revenue from say, no sales (i.e. revenue = £0) and sales of a quantity of 30 000 (i.e. revenue = 30 000 × £1.50 = £45 000).

A line can be drawn for the fixed costs, i.e. at £12 500.

The variable costs will be a straight line graph. To obtain two points determine from say, no quantity manufactured (i.e. costs = £0) and a production quantity of 30 000 (i.e. costs = 30 000 × £0.75 = £22 500).

The total costs are the sum of fixed costs and variable costs.

The break-even point, which is the junction of the graphs of revenue from sales and total costs, determines the quantity which has to be produced before the company starts to show a profit on its investment, i.e. in this case 16 700 components (see Fig. 1.27).

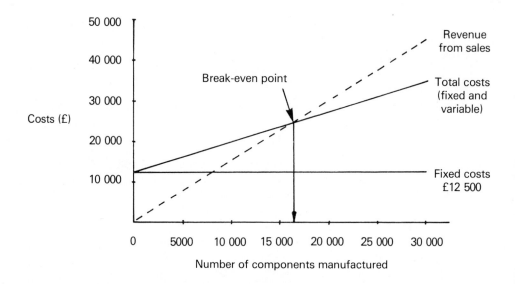

Fig. 1.27   A break-even chart.

Summary

Direct Expenses	Overhead Expenses

Costs incurred from: *Costs incurred for:*

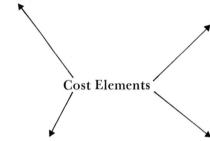

manufacture of special tooling
manufacture of packaging
work subcontracted to other firms

indirect materials
indirect labour
servicing
telephone
rent and rates
electricity
gas
warehousing
transport
advertising

Cost Elements

Direct Labour Costs

Wages paid to operatives who
are manufacturing the products

Direct Material Costs

Cost of material from which
the product is actually
manufactured

QUESTIONS

1. A steel bridge 250 m long is designed to operate within a temperature range of $-10°C$ to $38°C$. What is the difference between minimum and maximum lengths of the bridge? The coefficient of linear expansion for steel is $11 \times 10^{-6}/°C$.

2. An aluminium piston is fitted into a cast iron cylinder with a clearance of 0.015 mm on diameter at the working temperature of $345°C$. The diameter of the cylinder is 90 mm at $20°C$. Calculate the diameter of the piston at working temperature.

3. A brass collar, which has a bore of 35 mm, is to be shrunk on to a mild steel spindle, 35.15 mm diameter. Both sizes being measured at $20°C$. Calculate the temperature to which the brass collar must be heated to give a clearance of 0.025 mm all around the spindle.

4. A cast iron block measuring $25 \, mm \times 10 \, mm \times 15 \, mm$ at $20°C$ is to be used in a refrigeration system when its temperature will be reduced to $-25°C$. Calculate the reduction in volume.

5. An aluminium ball, 55 mm diameter at $25°C$, is heated to $135°C$. Calculate the new volume.

6. 50 ml of mercury, at a temperature of 125°C, are reduced until the volume is 49 ml. What is the temperature of the mercury after this reduction? Coefficient of volumetric expansion of mercury is $18 \times 10^{-5}/°C$.

7. Two flat plates are held together by three 10 mm rivets in a single lap joint. If the rivets shear when the plates are pulled apart by a force of 25 kN, calculate the maximum shear stress of the rivet material. If the tensile force on the plates does not normally exceed 5 kN, determine the factor of safety used.

8. In a small engine the connecting rod has a cross-sectional area of 150 mm². The piston diameter is 90 mm and the maximum explosive pressure is 2.5 MN/m². Determine the maximum compressive stress in the connecting rod.

9. Printed circuit boards are produced from laminates. Describe how their physical properties affect the choice of this material.

10. A pressure vessel has an internal diameter of 900 mm and is subject to a maximum internal pressure of 500 kN/m². Cover plates are secured on each end of the vessel by 12 high tensile bolts of ultimate tensile stress 875 MN/m². Determine the minimum diameter of the bolts using a factor of safety of 12.

11. A piercing tool is to punch out 5 holes, each 5 mm diameter, through 1 mm thick cold drawn mild steel strip. If the shearing strength of the material is 350 MN/m², find the force required for piercing and determine the compressive stress in each tool.

12. Containers are made from glass and plastic. Explain how the mechanical properties of each affect the choice of material.

13. State the two contributing factors that influence the rate of corrosion.

14. The corrosion process may be likened to the electrolytic action that takes place in a simple cell. Briefly describe this action, and state also the three essential components of this cell.

15. When a component is designed that is likely to operate in a corrosive environment, why is it undesirable to place dissimilar metals together, which are widely spaced in the electrochemical series?

16. Describe the underlying principle of sacrificial cathodic protection and gives two examples of its use.

17. With the aid of simple diagrams describe the principle of:
    (i) electroplating and (ii) anodising.

18. State the importance of surface preparation prior to the application of a coating, and briefly describe three methods of surface preparation.

19. Using simple sketches, give three examples to show how corrosion may be minimised by careful design.

20. Give two reasons why the selling price of a product must bear a fair share of the costs of production.

21. Briefly describe 'tooling costs'.

    Give one advantage to be gained by investment in tooling and its consequent volume production.

22. Draw a break-even chart and determine the break-even point for the production of a component with a selling price of £72.50 each, fixed costs of £110 000 and variable costs of £50.00 each.

ASSIGNMENT ———————————————————————————

It is necessary to select various materials for the manufacture of the following components:

1. jaw inserts for a bench vice;

2. vertical supports for the lifting device, shown in Fig. 1.28, which is to be used outside in a factory yard to load lorries;

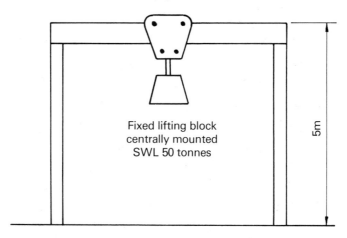

Fig. 1.28 Vertical supports with lifting block for assignment.

3. motor car bumper;

4. thermocouple;

5. double-glazing window section, as shown in Fig. 1.29;

6. housing for a retractable 2 m steel tape.

Suggest materials from which these components could be made giving reasons for your choice.

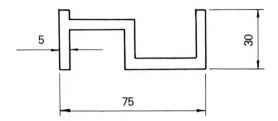

Fig. 1.29 Double-glazing window section for assignment

Your answers should include stress calculations where appropriate and comments upon:

(a) physical properties, e.g. melting point, density, etc;

(b) mechanical properties, e.g. hardness, ductility, etc;

(c) primary forming processes, e.g. casting, forging, etc;

(d) corrosion protection;

(e) cost properties for alternative materials or methods of production.

MANUFACTURING CONSIDERATIONS

The Comparison of Basic Manufacturing Processes

After reading this chapter you should be able to:

★ compare the basic manufacturing process for components with regard to cost and production problems (G);

★ compare the significant factors in the manufacturing processes of bending, forging, extrusion, welding, casting and plastics moulding (S);

★ give examples of how these processes and economic factors could influence the design (S).

(G) = general TEC objective
(S) = specific TEC objective

INTRODUCTION

In Chapter 1, certain factors were considered relating to the choice of material from which an article should be made. These factors were shown in Fig. 1.1.

Unfortunately these factors cannot be considered completely in isolation from one another as they are closely inter-related, i.e. one factor influences another. For example, the shape of a component, its size, its dimensional accuracy and the mechanical and physical properties of the material from which the component is made, will influence the choice of manufacturing process. Consider the manufacture of a motor car carburettor. This instrument is very intricate in shape, being made to a high degree of dimensional accuracy in a corrosion resistant zinc base alloy. Because of these factors, the most suitable manufacturing process used to produce the carburettor, other than lengthy machining from solid which would be impractical and expensive due to its complexity, is pressure die-casting. In turn pressure die-casting is only suitable for zinc and aluminium base alloys.

Consideration should also be given to the number of components to be produced. In the case of the pressure die-casting process, this process is only used for the production of large quantities of components as the cost of the tooling is very high.

The cost of tooling however is partly offset by the very accurate casting produced, which needs little or no subsequent machining.

The following chapter will examine briefly various basic manufacturing processes, together with the significant factors which may influence the choice of these processes.

BENDING ————————————————

Both ferrous and non-ferrous metals may be bent cold provided they are sufficiently ductile. If they are not of a particularly ductile nature, they may have to be heated until they become sufficiently plastic. This makes bending easier and also prevents the metal from cracking along the bend.

Assuming that the material is ductile and the design of the component not over complex, such that it cannot be bent, then the method used for bending will depend upon: (i) the thickness of the material to be bent; (ii) the length of the bend; (iii) the number of components to be bent.

Bearing these factors in mind, let us briefly examine the various ways in which metal can be bent.

Bending by Hand

Simple forming or bending of sheet metal up to 1.6 mm (16 gauge) may be carried out using one of the following hand methods.

   (i) If a simple, straight, angular bend is required, and the length of the bend is small, then the metal may be gripped in a bench vice and bent using a wooden or rawhide mallet.

   For longer bends the metal may be clamped between two pieces of wood, or metal, and bent.

   (ii) For other bent and formed shapes, bench stakes may be used. These are held in a bench plate that is fastened to the bench. The metal is worked over the bench stake using a wooden or rawhide mallet. Care must be taken not to damage or bruise the metal when hitting it with the mallet.

   Some common types of bench stakes, and their uses, are shown in Fig. 2.1.

Bending Using the Folding Machine

Larger sheets of metal, up to for example 2 mm (14 gauge) mild steel, may be bent using a hand-operated folding machine of the type shown in Fig. 2.2. Sheet metal up to 4 mm in thickness may be bent using a power-operated folding machine.

The two bending methods described are usually undertaken by skilled craftsmen and often the operations are time-consuming. For these reasons articles produced in this way are limited to very small numbers, perhaps only one or two off.

Typical articles produced in this way are, for example, hoods and ducts for air-conditioning and ventilating systems.

Hatchet stake

Used for making
sharp bends
and tucking in
wired edges

Bick iron

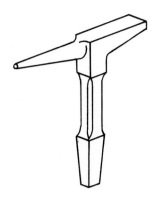

Used for
making spouts
and sharp
tapering
articles

Half-moon stake

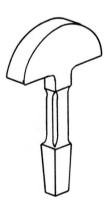

Used for carrying
out similar operations to
the hatchet stake
but on curved
surfaces

Side stake

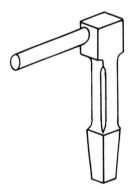

Used for
supporting
cylindrically
shaped articles,
e.g. seaming
pipe work

Fig. 2.1 Common types of bench stake.

Fig. 2.2   A hand-operated folding machine.   *(By courtesy of Walton of Radcliffe, Manchester.)*

Bending Using the Press Brake

This versatile machine may be used for bending and forming a wide variety of metals ranging in thickness from 1 to 12 mm. Because of the nature of the press brake's design, bends of considerable length can be made.

Press brakes may be operated by hydraulic or mechanical means.

Mechanical press brakes can be of a simple single-speed nature or have a more complex variable-speed system that offers not only fast speed and slow speed but rapid approach speed, slow bending speed and rapid return speed.

These multi-speed press brakes may be used for one-off components or high volume production of simple bent components. Fig. 2.3 shows a modern large-capacity mechanical press brake.

The hydraulically operated press brake is slow in action but is capable of greater bending pressures than mechanical press brakes.

The metal is bent or formed to shape between a fixed lower die and a moving upper punch, usually by air-bending or bottom-bending.

Air-bending

In air-bending (Fig. 2.4) the punch and die are relieved to permit air between the tools and component at all times.

Fig. 2.3 A modern large-capacity mechanical press brake. *(By courtesy of A Kinghorn and Co Ltd., Todmorden.)*

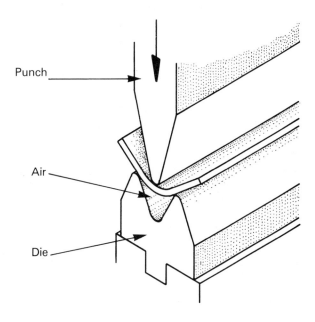

Fig. 2.4 Air-bending using the press brake. The desired angle of bend is obtained by the depth that the punch enters the die opening. *(By courtesy of A Kinghorn and Co Ltd., Todmorden.)*

The advantage of air-bending is the ability to bend various angles with the same tooling. The further the punch enters the die, the smaller will be the angle of the bend. Allowance, however, must be made for the 'spring back' of the metal. This is achieved by over-bending the metal such that when it is removed from the press the metal springs open to the correct angle. Unfortunately, because of the method of bending, sharp radii cannot be produced.

Bottom-bending

In bottom-bending (Fig. 2.5) the punch and die match exactly, so that tight inside-bend radii may be produced. To achieve this, the component is coined between the punch and die, i.e. the bend is produced by crushing the metal between the punch and die. This results in the bend retaining the angles of the punch and die.

This method requires at least three times the pressure to produce a given bend that the air bending method requires, but a much sharper radius can be produced. Bottom-bending is usually performed on hydraulic press brakes.

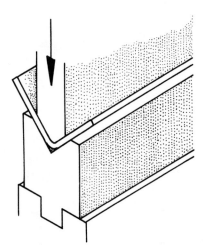

Fig. 2.5 Bottom-bending using the press brake. The angle on punch and die determines the angle of bend. *(By courtesy of A Kinghorn and Co Ltd., Todmorden.)*

Bending Pressure

The pressures required for bending are influenced by the type of material being bent and its thickness.

For example, with air-bending, taking the bending pressure of mild steel to be approximately $434\,MN/m^2$, the approximate bending pressures for the following materials will be:

Hard aluminium	50% mild steel pressure
Soft brass	60% mild steel pressure
EN 58 stainless steel	145% mild steel pressure

The capacities of press brakes vary considerably, and it is important to check, particularly when bottom-bending, that:

(i) the machine has sufficient power to produce the desired bend; and

(ii) producing the desired bend does not structurally damage the machine and tooling.

Press Tools

The production costs of large numbers of similar sheet metal articles can be reduced considerably by the use of press tools. In fact many of the mass-produced articles in everyday use, from cooking utensils to motor car bodies, will have been produced using press tools.

There is virtually no limit to the size and shape of components that can be produced on press tools provided that: (i) the number of components to be produced is sufficiently large to justify the high cost of tooling, and (ii) the press on which the tooling is to be used is of sufficiently large capacity.

Fig. 2.6 shows a simple pierce-and-blank 'follow-on' type of press tool used for the manufacture of washers.

A wide variety of operations may be undertaken using press tools. These may be classified as *bending, forming, piercing* and *blanking,* and are shown in Fig. 2.7. These operations may be carried out separately on individual press tools or more than one operation, such as pierce and blank, or pierce, bend and cut-off, may be incorporated in a single press tool, known as a 'follow-on' press tool. It must be realised however that these tools are more complex and therefore more expensive.

Modern machining techniques, such as the spark erosion process (in which metal is removed or eroded by an electrical discharge across a small gap between an electrode and the workpiece) enable shapes to be produced in the dies which could not easily be produced by conventional machining methods — thus enabling quite complex articles to be produced by press tools.

Mechanical presses are generally used with press tools, and their operating cycle can be very rapid. Together with the use of automated feeding of the metal into the press tool this enables very high production rates to be achieved.

Fig. 2.6 A press tool used for the manufacture of washers.

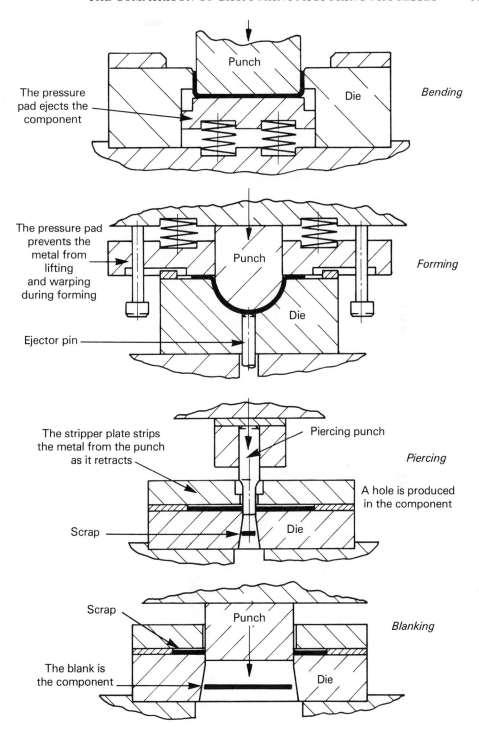

Fig. 2.7   Basic press tool operations.

Design Consideration When Bending

Bend Allowance

When a piece of material is bent, the material on the outside of the bend is stretched whilst that on the inside is compressed. When calculating developed lengths of components an allowance must be made for these effects.

At some intermediate or neutral position the material is neither stretched or compressed. Its length remains unaltered. Material on this neutral position is said to lie on a neutral surface (Fig. 2.8).

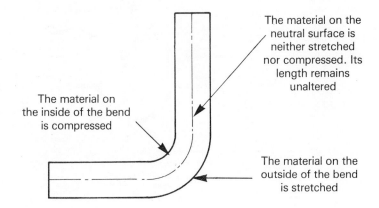

Fig. 2.8   The neutral surface.

Often the neutral surface is taken as the centre line of the material. Consider the following example.

Example   Taking the neutral surface to be along the centre line of the material calculate the length of the blank to form the clip shown in Fig. 2.9.

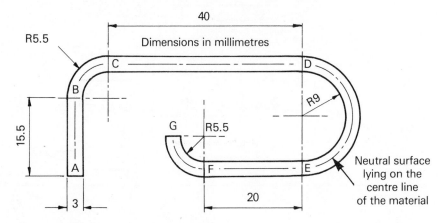

Fig. 2.9   Clip featured in Example.

Solution   The length required will be the length of the neutral surface:

$$\text{Length of neutral surface} = \underset{\text{(i)}}{AB} + \underset{\text{(ii)}}{BC} + \underset{\text{(iii)}}{CD} + \underset{\text{(iv)}}{DE} + \underset{\text{(v)}}{EF} + \underset{\text{(vi)}}{FG}$$

Calculate the individual lengths:

(i)   Length of AB = 15.5 mm

(ii) Length of quarter-circle BC. The radius to the neutral surface will be:

5.5 mm + 1.5 mm (half metal thickness) = 7.0 mm

Therefore the diameter will be 14.0 mm.

$$BC = \frac{\pi d}{4}$$

$$= \frac{22 \times 14}{7 \times 4}$$

$$= 11 \, mm$$

(iii)   Length of CD = 40 mm

(iv) Length of half-circle DE. The radius to the neutral surface will be:

9 mm + 1.5 mm (half metal thickness) = 10.5 mm

Therefore the diameter will be 21 mm.

$$DE = \frac{\pi d}{2}$$

$$= \frac{22 \times 21}{7 \times 2}$$

$$= 33 \, mm$$

(v)   Length of EF = 20 mm

(vi) Length of quarter-circle FG. This has the same radius as the quarter circle BC and will therefore be the same length.

$$\therefore \; FG = BC$$

$$= 11 \, mm$$

Total length of neutral surface = 15.5 + 11 + 40 + 33 + 20 + 11

$$= 130.5 \, mm$$

It should be noted however that the position of the neutral surface can alter depending upon the sharpness of the bend and the type, shape and thickness of material. Before bending takes place the position of the neutral surface should be calculated or obtained from Standard Tables. Alternatively a trial-and-error technique may be adopted, using scrap material.

The Effects of Bending

As we have seen earlier, material on the outside of a bend is stretched while that on the inside is compressed. The effect of this stretching and compressing of the material is to set up opposing tensile and compressive stresses. Where the effect of these stresses may affect the life and performance of the component, the metal should be either stress-relief-annealed after bending or bent hot. It is advisable to hot-bend materials such as high-carbon steels.

Materials above 3 mm in thickness that have been work hardened by rolling or drawing, etc. such as bright-rolled mild steel, should be annealed prior to bending. Without annealing these materials may crack along the outside of the bend, particularly if the bend is made parallel to the grain flow. Alternatively a material that is not in a work-hardened condition may be used, such as hot-rolled black stock.

Because of the elastic nature of cold steel there is a tendency for it to spring back slightly after bending. This is overcome by overbending the steel, and the extent of overbending must be taken into account when designing and manufacturing press tools.

When certain metals are bent or formed, e.g. aluminium and stainless steel, the tools may mark the metal. If the surface finish of the metal after bending is of prime importance, it could be coated with a thin plastic film, which can be peeled away after bending, leaving the metal unmarked.

Radii

When bending, using press tools, the radius of the bend must be considered. If the radius is too small, the metal around the bend will become distorted (Fig. 2.10(a)).

In general the radius on the inside of the bend should be no smaller than $1\frac{1}{2}$ times the material thickness t (Fig. 2.10(b)).

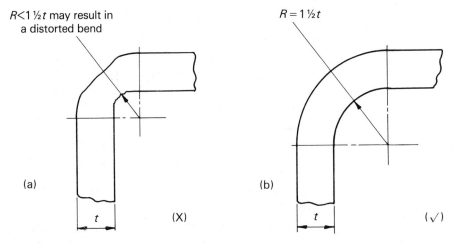

R<1½*t* may result in a distorted bend

$R = 1\frac{1}{2}t$

(a)

(b)

t (X)

t (√)

Fig. 2.10   Good (√) and bad (X) designs for bending.

Holes

When piercing holes, using press tools, the hole diameter should be no smaller than the thickness of the material t being pierced, if damage to the punch is to be avoided (Fig. 2.11).

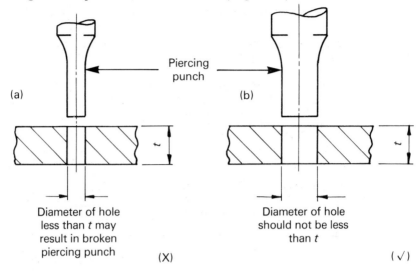

Fig. 2.11   Good ($\checkmark$) and bad (X) designs for piercing holes.

FORGING

Certain metals, such as high-carbon steel, are difficult to work cold because they are not sufficiently malleable or plastic.

Forging is a process whereby the metal is heated until it becomes sufficiently plastic to be forced into shape by hammering or squeezing.

There are several different methods of forging but the following advantages are common to all methods.

(i)   There is very little material wastage. The metal is worked almost to its finished shape, giving a reduction in the subsequent machining time compared with machining the component from solid.

(ii) Because of the plastic deformation that takes place during forging the grain flow will follow the shape of the component, giving it greater strength and resistance to shock-loading; Fig. 2.12(a) shows the grain flow in a forged component together with that of a similar component machined from solid at (b).

(iii) Working of the metal compacts the structure and reduces the grain size thus increasing the toughness and strength of the metal.

(a)

(b)

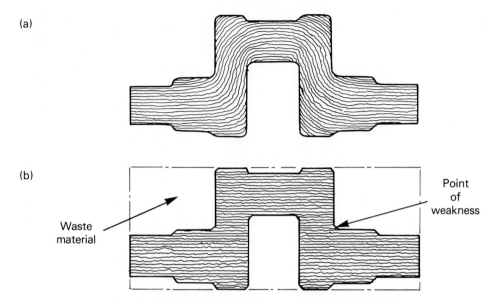

Waste
material

Point
of
weakness

Fig. 2.12   (a) The grain flow in forged components. The grain flow follows the shape
of the component giving greater strength; the component also requires a
minimum of machining.   (b) The grain flow of a similar component
machined from a solid blank of drawn material follows the direction of the
drawing. A reduction in strength may arise from the sudden change in
section. Also there is a large proportion of waste material.

Fig. 2.13 shows some typical forged components. The chosen method
of forging will depend upon the size, shape, mass and number of
components to be produced.

Hand Forging

This process uses the skills of a blacksmith to work the heated metal
to shape using hand tools. It is a time-consuming process and is used
only for the manufacture of very small numbers of components, in
plain carbon or alloy steels, which are of such a size that they can be
comfortably held and worked by hand. The basic operations of hand
forging are shown in Fig. 2.14 and may be described as:

(a) *drawing down,* whereby the metal is increased in length and its
cross section is decreased;

(b) *setting down,* which is similar to drawing down but the metal is
worked from one side only;

(c) *upsetting,* in which the metal's cross-sectional area is increased
and its length decreased;

Fig. 2.13 Examples of forged components. *(By courtesy of Doncasters Sheffield Ltd.)*

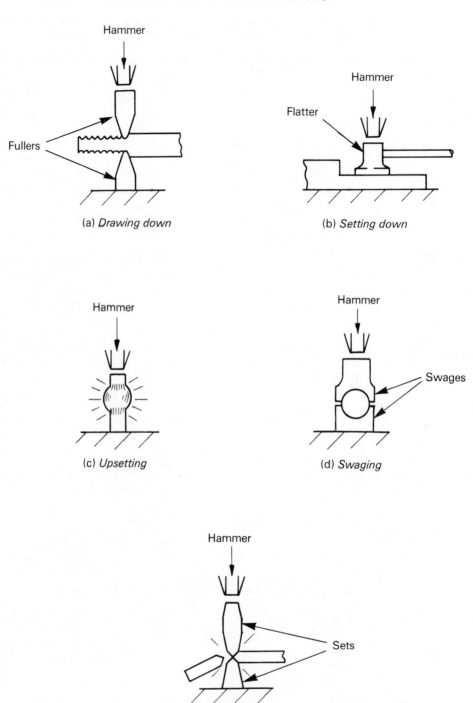

Fig. 2.14 Basic hand forging operations.

(d) *swaging,* which produces various cross-sectional shapes using tools known as swages;

(e) *cutting,* in which hot metal is cut and marked using tools, similar to cold chisels, known as sets.

Open-tool Forging

Open-tool or hammer forging consists of gradually working a heated metal ingot, which may be manipulated by hand or mechanical means, between a fixed bottom anvil and an upper moving hammer. The hammer may be operated hydraulically or by steam.

The operations carried out by open-tool forging are basically the same as those in hand forging: drawing down, upsetting and swaging.

Open-tool forging is not a rapid process but because of the simple tooling used and the ability to produce forgings ranging from a few kilograms to several tonnes, in materials such as high carbon steel, alloy and stainless steel, it is an economical process for low production runs.

Open-tool forging improves the soundness of the component by welding up any cavities or porosity that have been formed during the solidification of the raw ingot.

Typical forging produced by this process range from bolts, bars, rings, shear blades and crankshafts to large turbine rotors for electricity generation.

This process is also used as a preparatory operation to drop forging to bring the component to a size approximating that given to it by the drop forging dies.

Drop Forging

Drop forging is used to produce large quantities of identical components in a wide variety of metals, such as carbon and alloy steels, aluminium, magnesium, copper, nickel and titanium alloys.

A heated metal billet is placed between a fixed lower die and a moving upper die, which is allowed to fall onto the billet of metal. The metal, being in a plastic state, is forced into the die cavity. Any excess metal is allowed to flow into the flash gutter (Fig. 1.26(f)); the flash produced is removed from the finished forging later.

Components that are simple in shape can be produced by repeated blows using one pair of dies; but several pairs of dies may be required if the component is more complicated in shape.

The cost of producing drop forging dies is high and must be justified by the production of large quantities of components.

Drop-forged components can be produced to a high degree of accuracy, very quickly, and require a minimum of machining to bring them down to precise dimensions.

Typical components produced by drop forging are connecting rods, crankshafts, stub axles, lifting tackle and heavy-duty components for agricultural, mining and earth-moving machinery.

Design Considerations When Forging

Hand and Open-tool Forging

Bearing in mind that the metal has to be manipulated by hand, and also the relatively simple tooling that is used, the designer should avoid over-complicated shapes, thus allowing easier forging and greater economy. Some examples of satisfactory and unsatisfactory design are shown in Fig. 2.15.

Drop Forging

There is virtually no limit to the shapes that can be drop forged, although consideration must be given to the following:

(i) In order to facilitate removal of the finished forging from the die it is necessary for the component to have a slight taper or draft (Fig. 2.16), usually between 0° and 10°, depending upon the design of the component, the material from which it is made and the type of forging equipment used.

(ii) The line along which the two halves of the die meet is called the parting line (Fig. 2.16). The variables that influence the location of the parting line are:

(a) the type and size of the forging equipment used;

(b) the shape of the component to be forged;

(c) the procedure for removing the flash;

(d) the grain flow;

(e) material consideration;

(f) the position of locating surfaces for subsequent machining.

Depending on the requirements, the parting line may be straight or irregular and complex. It is preferable to keep the parting line positioned in one plane whenever possible. This makes die manufacture, forging and subsequent flash removal operations simpler, and therefore less costly.

(iii) The designer should consider the possibility of avoiding right- and left-handed components, thus reducing the cost of the dies considerably. By careful design one component can be made to serve both right- and left-handed applications.

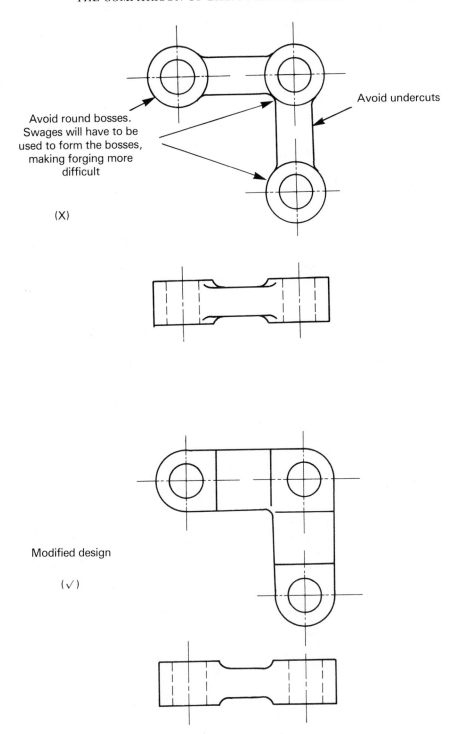

Avoid undercuts

Avoid round bosses.
Swages will have to be
used to form the bosses,
making forging more
difficult

(X)

Modified design

(✓)

Fig. 2.15 Simple design considerations when hand and open-tool forging *(continued overleaf).*

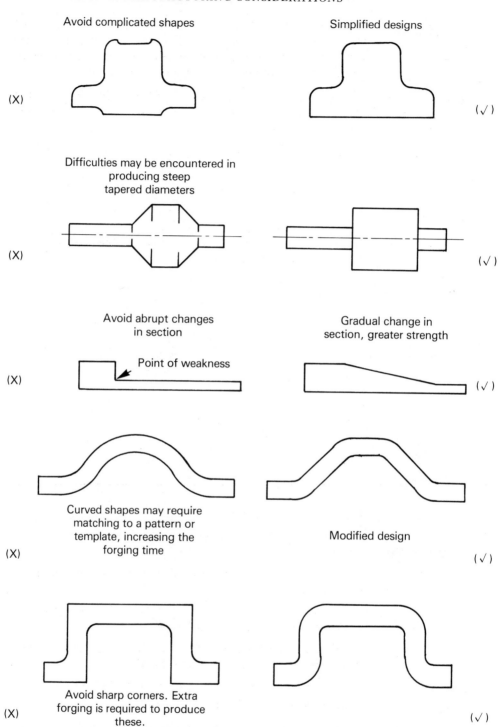

Fig. 2.15 *(continued)*

(iv) Webs and sections should not be made too thin (Fig. 2.16) because they will cool very quickly, resulting in greater forging pressures being required. As a result, greater stresses are imposed on the dies and forging equipment.

(v) Generous corner radii (Fig. 2.16) should be provided wherever possible. This will assist in metal flow and give improved die life.

Allow generous radii where possible

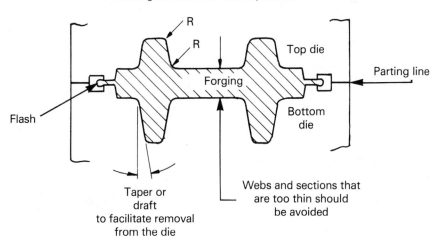

Fig. 2.16   Design considerations when drop forging.

EXTRUSION

Extrusion is a process by which long lengths of various cross-sections can be produced very quickly by squeezing metal through a steel die (Fig. 1.26(e)).

Extrusion is carried out either hot or cold depending upon the type of metal, for example:

(i) Lead, tin and their alloys can be extruded hot or cold.

(ii) Copper and brass are extruded hot.

(iii) Aluminium, magnesium and their alloys are extruded hot or cold depending upon the type of alloy.

(iv) Plain carbon and alloy steels are extruded hot.

(v) Nickel base alloys are extruded hot.

An unlimited variety of cross-sections can be extruded, including undercuts. Extrusions can be designed for assembly to each other by dovetail and sliding joints. Fig. 2.17 shows examples of various extruded sections.

Fig. 2.17 Various examples of extruded section in copper-based alloys. *(By courtesy of Delta Extruded Metals Co. Ltd.)*

Since plastic deformation takes place during extrusion the product will exhibit a definite grain flow, in the direction of extrusion, giving an improvement in strength. Cold extrusion achieves a fine surface finish together with a close dimensional accuracy, requiring little or no subsequent machining, although the metal itself may be in a severely work-hardened condition.

For their size, extrusion dies are expensive and this cost must be justified by large quantity production.

Extruded sections have many applications in both mechanical and civil engineering, including curtain rails, double-glazing window sections, automobile radiator grills, structural members for aircraft, buildings and ships. Fig. 2.18 shows various uses for extruded section.

Design Considerations When Using Extrusions

Extrusions can be obtained in an almost unlimited variety of shapes but, as stated earlier, the cost of producing the dies is high and must

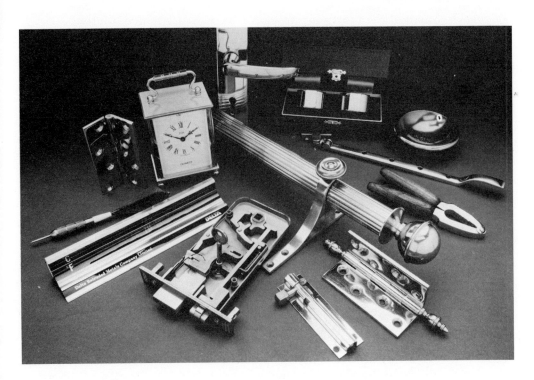

Fig. 2.18 Products of extruded sections: (a) brass turned parts; (b) finished products. *(By courtesy of Delta Extruded Metals Co. Ltd.)*

be justified by large quantity production. If a small batch quantity of components is to be produced from extruded material, economies may be made by using standard stock section. Consequently the designer should consult the extrusion manufacturer before finalisation of the design.

WELDING

In welding the joint faces of the pieces of metal to be joined are heated, by various methods, which causes their surfaces to be converted into a liquid state and allows them to fuse together. A filler metal, with a similar melting temperature to that of the parent metal, may or may not be used.

There are many welding processes and to explore each of them is beyond the scope of this chapter. A large number of the processes can be dealt with under three main headings:

(i) gas welding;

(ii) arc welding;

(iii) resistance welding.

Some of the remaining processes are also mentioned briefly.

Gas Welding

This process basically consists of heating the metals to be joined, until they fuse together, by means of a flame which is produced by the combustion of a gaseous mixture. A filler material may be used depending upon the metals and their thicknesses.

Gas welding can be undertaken on a jobbing basis and is very economical because of the low cost of the necessary equipment.

The most widely used gas welding method is oxyacetylene welding. By varying the ratio of oxygen and acetylene both ferrous and non-ferrous metals may be welded.

Arc Welding

An electric arc is formed when an electric current passes through the small air gap separating two metal electrodes. In arc welding one electrode is the welding rod and the other is the metal to be welded. An arc is formed between the very small gap left between them and the resulting electrical energy is converted into heat and light energy and a temperature is developed which is sufficiently high enough to melt the parent metal. We now consider four types of arc welding.

Manual Metal Arc Welding

Manual metal arc welding is the most common form of arc welding. It is a versatile process which is easy to set up and the capital cost of equipment is low.

The metal core of the electrode passes the electrical energy to the arc and is melted together with the flux coating to form molten droplets of flux and metal. High-temperature gases are obtained from the flux coatings, which provide a shield, and protect the molten metal at the weld face from the ingress of the atmosphere (see Fig. 2.19).

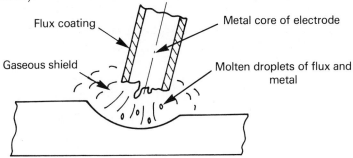

Fig. 2.19   Manual metal arc welding.

Tungsten Inert Gas Welding

Tungsten inert gas welding (TIG) is an arc process in which an arc is maintained between a non-consumable electrode and the workpiece. The weld area is protected from atmospheric contamination by a stream of inert gas, usually argon (see Fig. 2.20).

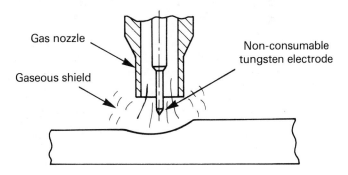

Fig. 2.20   Tungsten inert gas welding.

Metal Inert Gas Welding

Metal inert gas welding (MIG and CO_2 welding) is also an arc process which involves feeding a current-carrying wire on to the workpiece through a torch which carries a shielding gas similar to that used in the TIG process.

There are numerous combinations of wires and gases which can be used, the most common ones being aluminium wire and argon (MIG welding) and steel wire and carbon dioxide (CO_2 welding).

An advantage of this process is that it can cover the range from thin sheet to plate successfully and most metals can be welded.

Submerged Arc Welding

Submerged arc welding is a process, like metal inert gas welding, which also employs a current-carrying wire electrode. The weld metal is protected in this instance by flux which is fed from a hopper to cover the weld area totally, and no arc is visible (see Fig. 2.21).

Long lengths of joint are ideal applications for this process.

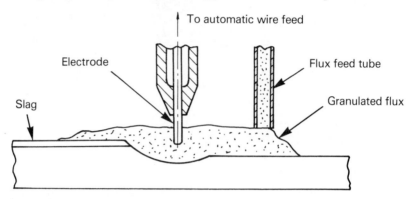

Fig. 2.21   Submerged arc welding.

Resistance Welding

If two or more metal parts are clamped between two copper electrodes, and a powerful electric current is passed through them, heating will occur in proportion to the electrical resistance of the circuit. The heating effect is greatest within the area where the parts are held in contact.

If the heat produced at these points is sufficient to attain the melting temperature of the parts and is combined with the force exerted by the electrodes, it will cause molecular interpenetration.

When the electrical and mechanical forces are removed and the metal has cooled, the slug of metal solidifies and forms a weld between the parts. There are five main types of resistance welding.

Spot Welding

Two truncated cone electrodes apply pressure to the parts to be welded and determine the area through which the current will flow to produce the spot weld (see Fig. 2.22).

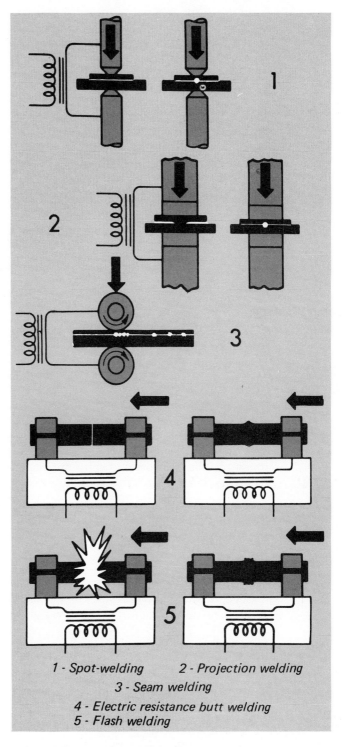

Fig. 2.22   Diagrams showing different processes for electric resistance welding *(By courtesy of Sciaky Ltd.)*

Spot welding machines have been used for a number of years by operators on motor car production lines, but it is a hard and laborious job. Latterly robot spot welders have been introduced which has removed the tedium of the work. The cost of these machines is extremely high, however, and long production runs are needed to justify the investment.

Miniature spot welders are used in the electronics industry. The work is light and women are usually found to be more dexterous in their use.

Projection Welding

Protrusions are pressed on one or more of the pieces of metal to be welded and give the location of the weld. When the current is passed through the electrodes the protrusions are compressed and the metal is welded together (see Fig. 2.22).

Typical applications are the automatic welding of brackets on to containers.

Seam Welding

The electrodes are in the form of wheels, which rotate either continuously or intermittently on the parts to be assembled. The current is passed in pulses and thus a row of spot welds is obtained (see Fig. 2.22).

Typical applications of seam welding are shown in Fig. 2.23.

Resistance Butt Welding

The two parts are held in clamps and brought together under pressure. The heat produced by the current brings the faces in contact to a plastic state and the applied pressure effects the weld (see Fig. 2.22).

Resistance butt welding is ideal for small compact sections as well as thin sheet.

Flash Butt Welding

The process differs from resistance butt welding in that voltage is applied to the workpieces before their ends are brought into contact. As soon as they touch, flashing takes place.

When the ends have been raised to the correct temperature, sudden forging pressure is applied and the current cut off (see Fig. 2.22).

Aluminium and its alloys, copper and its alloys and steel can be successfully flash-welded.

Friction Welding

Friction welding transforms mechanical energy into heat through rubbing two surfaces together under pressure. The two parts to be

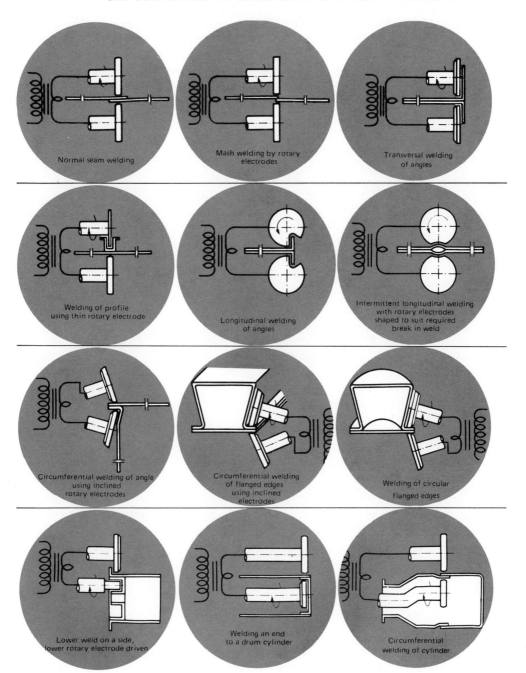

Fig. 2.23 Applications of seam welding. *(By courtesy of Sciaky Ltd.)*

joined end to end are held, one in a fixed clamp and the other in a rotating clamp. They are brought into contact under a predetermined pressure. The friction forces produce heat which rapidly raises the temperature of the parts to welding temperature. The rotation is stopped and an axial force is applied to form the weld (see Fig. 2.24).

This system of welding can be applied to:

    solid-to-solid sections;

    tubes to tubes;

    solids to tubular sections;

    tubes to flanges.

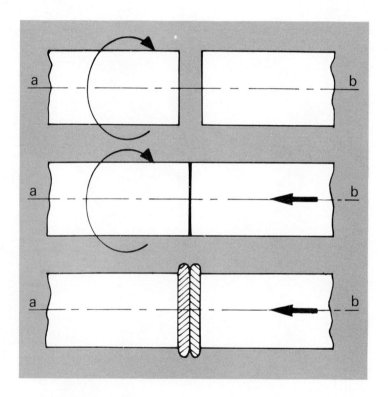

Fig. 2.24   Friction welding.   *(By courtesy of Sciaky Ltd.)*

Electron-beam Welding

Electron-beam welding is a process in which the heat required for fusion is derived from the kinetic energy of a dense beam of high-velocity electrons. This beam of electrons bombards, heats, melts and welds metals in a vacuum at high speed.

The workpiece is traversed under the beam by precision work-handling equipment which enables welds to be produced which are of high quality and at fairly high production rates. Because heat is generated at a very high rate across the face of the joint, only a minimum amount of metal necessary for fusion is melted producing hardly any distortion (see Fig. 2.25).

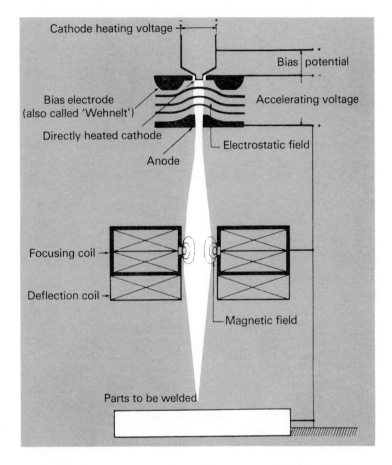

Fig. 2.25 Electron beam welding. *(By courtesy of Sciaky Ltd.)*

Design Considerations for Welding

Welding has the advantage of a great saving of material, low production costs and offers the designer greater scope in design. In the motor-car and aircraft industries the principle of welding has been a vital function in the development of lightweight methods of construction. One of the great advantages of welding is that the strength of the welded joint is virtually the same as that of the parent metal.

Some of the rules to follow when designing a component for welding are as follows.

    (i) When replacing a casting with a welded fabrication do not try to copy the casting exactly.

   (ii) Keep the number of welds to a minimum.

  (iii) Use location methods where possible to facilitate welding.

  (iv) Use butt welds wherever possible.

   (v) Try and make the ends to be welded of equal thickness.

Deciding between different processes, which provide equally acceptable welds, can lead to complicated economic evaluations which may require the help of the specialist. It may prove to be more economic to put the work out to a subcontractor.

Design, material and process cannot be selected independently, but generally design and material are considered first and then modified as the decision on process selection is developed.

Material thickness is very important, i.e. thick plate cannot be welded by the spot welding process. Joint shape is similarly important, i.e. steel plate cannot be friction-welded as the process demands circular or near-circular components.

For large numbers of welds consideration must be given to automatic methods which, when set, will consistently produce welds of higher quality than manual welding.

One of the most important points in all types of welding is the provision of adequate accessibility for the welding process. Accessibility enables the welder to produce sound welds and, conversely, inadequate accessibility could lead to unsound welds.

Welding labour time should be reduced to a minimum. It is more economic to bend plate than weld two plates at right angles. Fig. 2.26 shows the construction of box sections. Simple right-angle plate locations are made at (a) and (b) using fillet welds, but at (c) a vee butt weld is used to join two plates which have each been preformed into a semi-elliptic shape which produces a box of greater strength and reduces welding time.

Systems of location help the welder produce a more satisfactory weld joint. Fig. 2.27(a) shows a simple method of welding a plate ring on to a boss; at (b) a shoulder is used as a method of location which will help facilitate the task.

Localised plate reinforcement is useful and examples are shown in Fig. 2.28, where a boss is welded to a plate to help support a bolting system, and in Fig. 2.29 where increased thickness is required to house a ball or roller bearing.

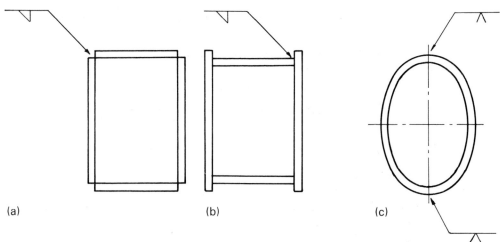

(a)   (b)   (c)

(a, b) Simple plate constructions using fillet welds

(c) Two plates preformed into a semi-elliptic shape and joined by vee butt weld to give a stronger box

Fig. 2.26   Construction in box sections.

Simple method of welding a plate ring to a boss

Alternative method using a shoulder to locate the plate ring and facilitate the task

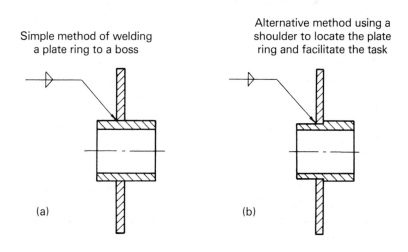

(a)   (b)

Fig. 2.27   The location method for welding.

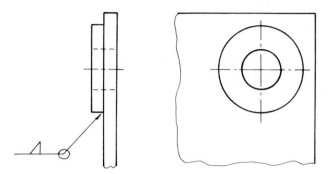

Fig. 2.28 Plate reinforcement: boss welded to plate to give localised support for bolting system.

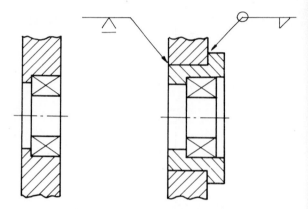

Fig. 2.29 Plate reinforcement: wall thickness increased locally to support ball or roller bearing.

Plates less than 3 mm thick can be butt-welded without any edge preparation (Fig. 2.30(a)) but when thicker plate is welded it is necessary to bevel the edges of the plates (see Fig. 2.30(b)).

(a)   (b)

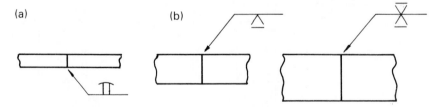

Fig. 2.30 Butt welds: (a) for thin plates (no preparation); (b) for thicker plates (bevel first).

Where possible welded joints should be of equal thickness. Where big differences in plate are involved (see Fig. 2.31) method (a) is unsatisfactory; method (b) will produce a better weld joint.

(a) (b)

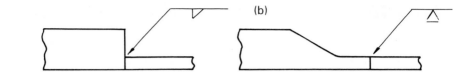

Fig. 2.31 Welding joints of different thickness: (b) is better than (a).

Placing several heavy welds together will cause internal stresses, and care should be taken so that welded seams do not interfere with each other (see Fig. 2.32).

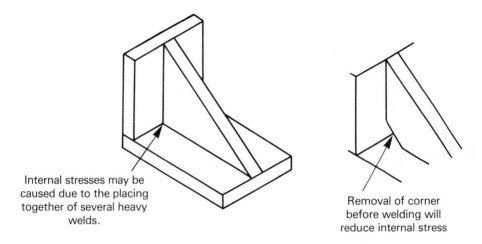

Internal stresses may be caused due to the placing together of several heavy welds.

Removal of corner before welding will reduce internal stress

Fig. 2.32 Reducing internal stresses.

Fig. 2.33 shows sample spot welds and areas where difficulty may occur.

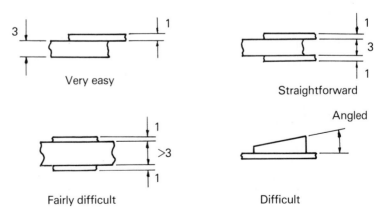

Very easy

Straightforward

Angled

Fairly difficult

Difficult

Fig. 2.33 Spot welds: a comparison of assemblies.

CASTING

This is the oldest and most basic of the metal working processes and is the shortest route between the raw material and the finished component. Casting consists of pouring or injecting molten metal into a mould which is exactly the shape of the required component, and although the system appears simple it requires knowledge and practical ability to obtain satisfactory results.

There are several methods of casting but in this chapter only sand-casting and pressure die-casting will be considered.

Sand-casting

A wood or metal pattern is made to represent the required part of the finished component. One half of the pattern is mounted on a board and the lower moulding box, the drag box, is placed in position surrounding the pattern (see Fig. 2.34(a)). Sand is rammed into the drag around the pattern and the upper surface strickled off level.

The complete drag, including its half-pattern, is inverted and the upper part of the pattern is mounted on to it. The upper moulding box, the cope box, is placed on top of the drag and tapered wooden pegs are placed in position, before ramming, to produce the runner and riser (see Fig. 2.34(b)).

The cope and drag are parted, the pattern and runner and rise pegs removed, and the surface of the mould coated with refractory dressing to prevent metal burning into the sand and to improve the casting finish. The ingate is cut and the core placed into position. The cope is located onto the drag and refractory bushes are mounted above the runner and riser positions (see Fig. 2.34(c)). The metal is then poured into the pouring bush until it appears at the top of the riser.

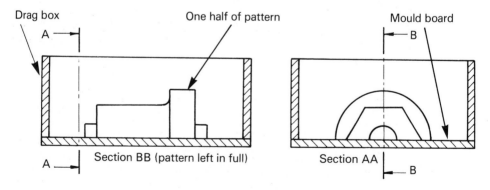

(a) One half of the pattern and the drag are mounted on a board ready for ramming.

Fig. 2.34   The sequence for the preparation of a mould *(continued opposite).*

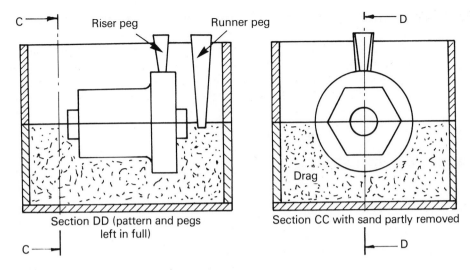

Section DD (pattern and pegs left in full)

Section CC with sand partly removed

(b) The drag is turned over and assembled with the other half of the pattern. The riser and runner pegs are in position. The cope is ready for ramming.

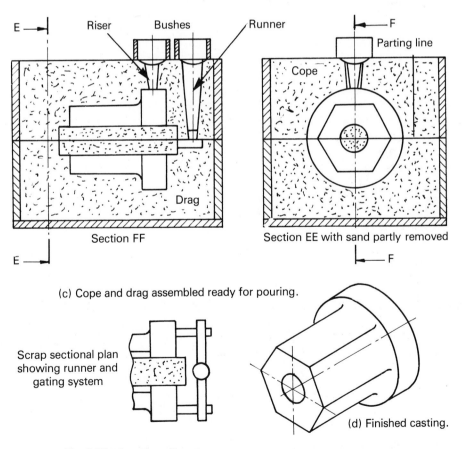

Section FF

Section EE with sand partly removed

(c) Cope and drag assembled ready for pouring.

Scrap sectional plan showing runner and gating system

(d) Finished casting.

Fig. 2.34 *(continued)*

The riser allows air to escape from the mould and acts as a reservoir allowing molten metal to feed back into the mould as cooling and contraction take place.

When the metal has solidified, the sand is broken away and the casting removed. The runner and riser are cut off and the casting is fettled to remove flash leaving the finished component, as shown in Fig. 2.34(d).

Pressure Die-casting

Pressure die-casting is one of the most important production processes for making non-ferrous components. Development is being undertaken to produce ferrous metal components, but systems capable of achieving this are not yet a marketable viability.

Pressure die-casting provides accurately dimensioned parts by forcing molten metal under pressure into a split metal die which subsequently opens to allow the parts to be ejected. Fig. 2.25 shows the various stages in producing a pressure die-casting.

Because of the expense in manufacturing dies, die-castings are used for large quantity production. The dies will retain their form over long periods of production during which many thousands of identical castings will have been made. Dies for zinc die-castings will have a life of up to a million components. However, there are many components which can be die-cast in much smaller numbers without the cost of the die making parts uneconomic: somewhere between 5000 and 20 000 components may be sufficient to amortise tooling costs in some cases.

Die-castings can be produced more rapidly than sand-casting with substantially lower labour costs and fewer secondary operations are required. Additionally die-casting can be produced with thinner walls, to closer dimensional accuracy, with smoother surfaces and more pleasing appearance.

Typical examples of components made by this process are shown in Fig. 2.36.

Design Considerations When Sand-casting

To ensure a proper flow of metal and filling up of the mould it is essential to round corners and slope surfaces. During casting, gas bubbles from the hot metal must be allowed to escape and this can be achieved by sloping the walls (see Fig. 2.37).

After the casting has solidified there will be further reduction in size as it cools down to the general foundry temperature. This shrinkage will depend upon the type of metal being cast and the pattern must be made correspondingly larger to allow for this reduction.

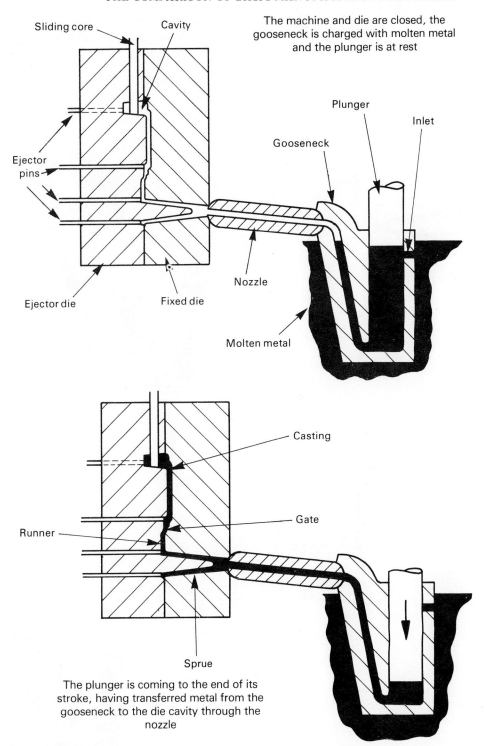

Fig. 2.35 Hot chamber die-casting *(continued overleaf).   (By courtesy of the Zinc Alloy Die Casters Association.)*

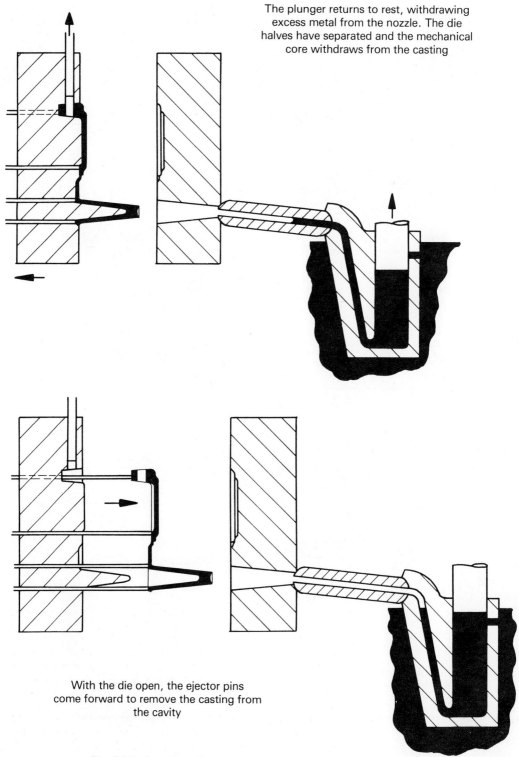

The plunger returns to rest, withdrawing excess metal from the nozzle. The die halves have separated and the mechanical core withdraws from the casting

With the die open, the ejector pins come forward to remove the casting from the cavity

Fig. 2.35   *(continued)*

(a)

(b)

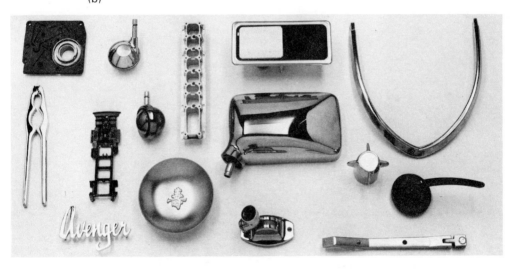

Fig. 2.36 Typical components made by the die-casting process. (a) A range of parts showing how features which would be difficult or expensive to make in any other way can be produced simply and accurately in zinc die-castings: internal and external threads, gears; combined where necessary with eccentric features and cams; and irregular shaped parts which made any other way would involve complicated and costly assembly work. (b) An indication of the range of finishes applied to zinc die-castings: chromium plating for motor car parts, hardware and bathroom fittings; black chrome for hardware; gold plate for decorative goods; silver plate for electronics; anodising for pneumatic motor parts; paint for toys; and lacquers for furniture fittings. *(By courtesy of Zinc Development Association.)*

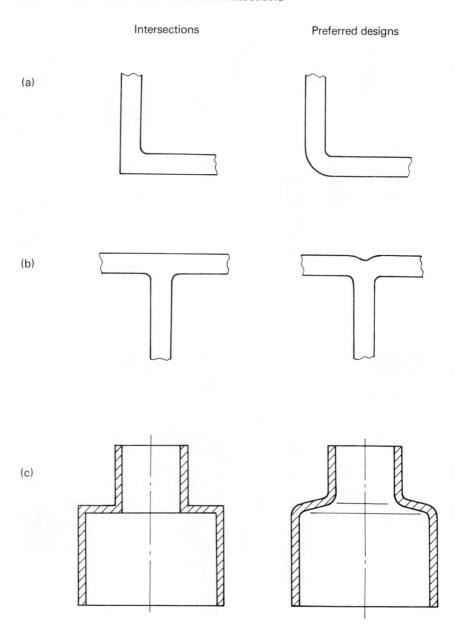

Intersections   Preferred designs

Fig. 2.37 Design of intersections.   (a) The square corner should be removed to give a uniform section or the curved portion could be made slightly thinner than the thickness of the wall.   (b) The intersection at a tee junction should be indented to reduce localised increased mass of metal. (c) Corners should be curved and walls sloped to allow gas bubbles to escape from the hot metal.

Sufficient metal must also be allowed on any surface which is to be machined afterwards.

As the casting solidifies the resulting contraction will set up stress in the metal and cavities will form which make the casting unsound. These cavities result from the fact that the outside of the casting and the thinner parts are the first to cool and solidify. Ideally uniformity of section should be made at the design stage. If this is not possible, then the transition from one section to another should be progressive (see Fig. 2.38).

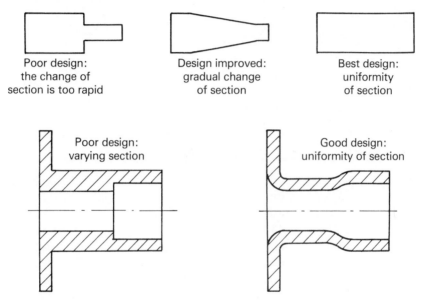

Poor design:
the change of
section is too rapid

Design improved:
gradual change
of section

Best design:
uniformity
of section

Poor design:
varying section

Good design:
uniformity of section

Fig. 2.38   The necessity for uniform sections.

When two ribs intersect at an angle, it is better to redesign so that the junction is perpendicular to reduce concentration of metal (see Fig. 2.39).

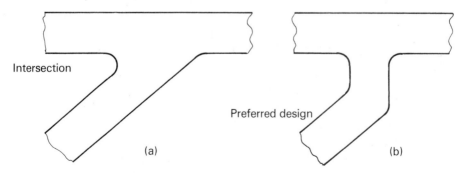

Intersection

Preferred design

(a)

(b)

Fig. 2.39   The intersection of ribs at an angle. The junction at (b) has better cooling
properties as the volume of metal there is smaller than at junction (a).

The volume of metal at junctions of ribs on a casting can be reduced by staggering (see Fig. 2.40).

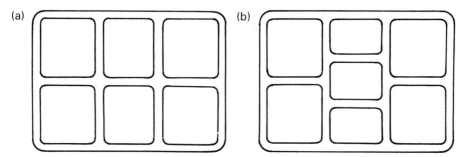

Fig. 2.40   The intersection of ribs at right angles. The intersection of ribs has been staggered at (b) to reduce any distortion in cooling.

Design Considerations When Die-casting

Because of the medium strength values of non-ferrous metals it is impracticable to produce parts which are highly stressed, hard or wear-resistant as die-casting. A practice used to increase these properties of a die-casting is to introduce into the casting parts which are made of more suitable materials. This is done by casting in inserts of more appropriate materials. For highly stressed internal threads a steel nut may be cast in position (see Fig. 2.41(a)). A phosphor-bronze bush may be used as a cast insert when a bearing location is required (see Fig. 2.41(b)). For highly stressed external threads a steel stud may be utilised (see Fig. 2.41(c)). Flats or knurled portions machined on the inserts prevent rotational or longitudinal movement.

As in sand-casting the rounding of corners is a vital aspect which must be undertaken to promote easier flow of metal along the thin wall sections. The walls should be of uniform thickness to prevent shrinkage cavities. A thinner section will solidify more rapidly than a thicker section and, due to the rigidity of the steel dies, shrinkage is more dangerous than that which occurs in sand-casting and may lead to cracking (see Fig. 2.42).

Cores are used to form holes in the casting and the ease by which these can be produced is one of the major advantages of this process.

Elimination of machining is another great benefit, and every effort should be made to take advantage of the accuracy of die-casting. Although extremely fine tolerances can be produced in die-castings, the finer the tolerance the greater the cost, because of the increased precision required in die making and maintenance. Designers should ask for tolerances no closer than are necessary.

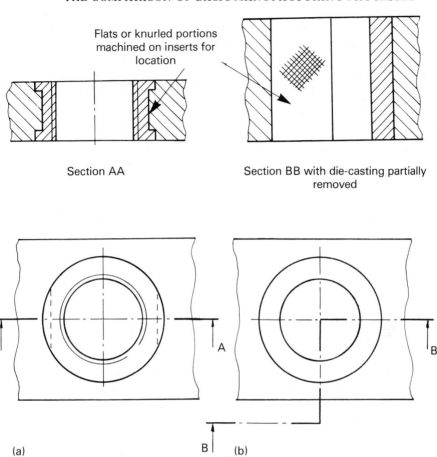

Flats or knurled portions machined on inserts for location

Section AA

Section BB with die-casting partially removed

A   A   B

(a)   B   (b)

(c)

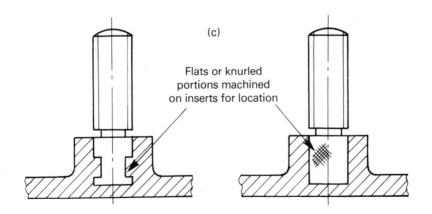

Flats or knurled portions machined on inserts for location

Fig. 2.41 Inserts in die-cast components: (a) steel nut; (b) phosphor-bronze bush; (c) steel stud.

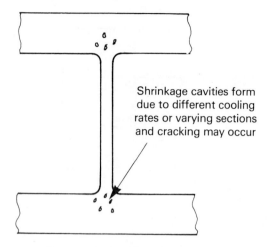

Fig. 2.42   Problems from non-uniform wall thickness.

Many die-castings are used in engineering applications where no surface treatment is required. However, in some circumstances coatings are necessary: to provide a decorative finish; to give outstanding corrosion resistance; and to produce an abrasion- and wear-resistant surface.

Finishes for Die-castings

The range of finishes for die-castings is very wide. Some of these are as follows.

Texturing

Although it is possible for the as-cast surface of a die-casting to provide a satisfactory base for any type of coating, it is generally better to use either a textured surface (see Fig. 2.43) or to allow some mechanical surface preparation before applying the finish. Where a broad surface is necessary on a die-casting texturing helps to avoid unsightly superficial blemishes.

Chemical Finishes

There are a number of chemical conversion treatments for zinc die-castings. In these processes chemical solutions react with the zinc surface converting it into a complex chemical layer with two principal effects – firstly it gives a degree of corrosion protection, and secondly is provides a good basis for subsequent paint or plastic coating. Chemical finishes used include *anodising*, which improves the abrasion resistance of zinc, and *phosphate coating* (see pp. 40-1).

Metallic Coatings

Zinc die-castings will accept many different electroplated finishes for decorative and protective purposes or to obtain special electrical or other surface properties (see Fig. 2.44).

(a)

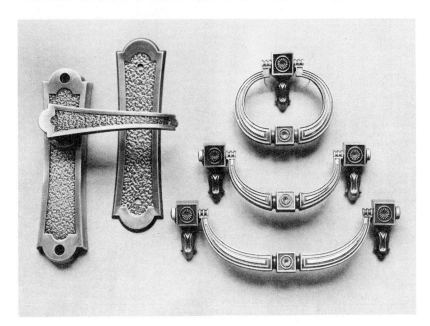

(b)

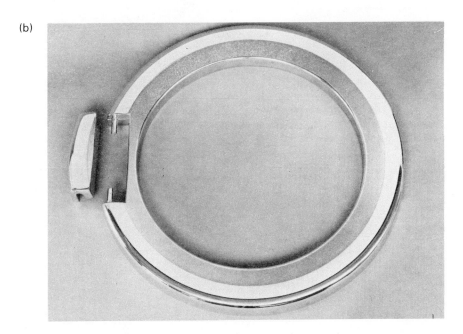

Fig. 2.43 (a) Some textured die-castings from Australia. Textures engraved onto the die are reproduced on all castings made from that die. *(By courtesy of Zinc Development Association.)*

(b) Textured zinc die-castings. Washing machine door bevel with textured finish on the inner ring portion. Copper/nickel/chromium plated. *(By courtesy of Zinc Development Association.)*

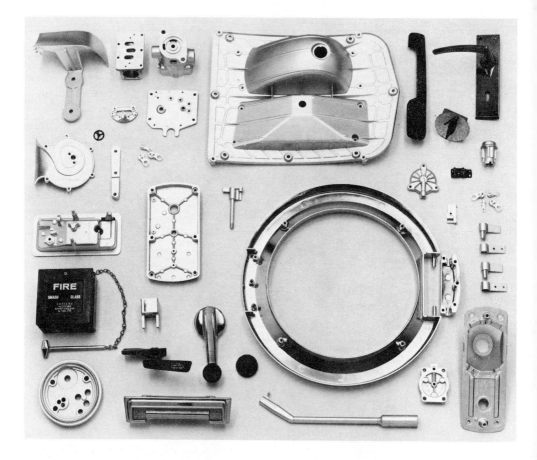

Fig. 2.44   A collection of zinc die-castings displaying various applied finishes – copper/
nickel/chromium plating, paints, powder coatings, and chemical black. The
car door handle at the bottom has a textured surface, engraved in the die.
(By courtesy of Zinc Development Association.)

Organic Finishes

(i) *Painting*. Many zinc die-castings are painted. Paints are often used in conjunction with plating for special visual effects (see Fig. 2.44).

(ii) *Plastic coating*. These powder coatings are being used increasingly for both decorative and protective finishes.

PLASTIC MOULDING ————————————————

The plastics industry can be said to have begun in the mid-nineteenth century when the manufacture of cellulose nitrate was evolved. However, it was not until the advent of the Second World War that the industry was given a tremendous stimulus. Phenol formaldehyde, commonly known as Bakelite, was moulded mainly to produce electrical components. Since that time several other moulding techniques have been developed, some of which are detailed below.

Since their introduction plastics have been used as cheap alternatives to metals. But since the increase in oil prices, in the latter part of the nineteen seventies, plastic is no longer a cheap material and designers must now look at plastics as engineering materials in their own right. The four types of plastic moulding are: compression moulding, transfer moulding, injection moulding and blow moulding. We shall consider each in turn.

Compression Moulding

This is used almost exclusively for thermosetting resins to produce components such as washing machine agitators, motor-car instrument panels, electrical plugs and switches, buttons and knobs.

The moulding compound in powder or pellet form is placed in a split mould. The mould is closed and the heat and pressure applied, gently at first to allow the compound to fill the mould, and then with full pressure (see Fig. 2.45).

Transfer Moulding

This is similar to compression moulding in that the compound is cured in a heated mould under pressure. It differs from compression moulding in that the moulding compound is heated until plastic in a cylinder, from which it is pushed by a ram through a series of runners and sprues into a heated split mould. Cycle times are faster than with compression moulding, but there is a large proportion of waste material since that which solidifies in the runners and sprues cannot be re-used (see Fig. 2.46).

Injection Moulding

This method of moulding is the one with the widest application. All thermoplastic materials can be moulded by this process and it is used to produce a wide range of components, a few of which are listed in Table 2.1.

Moulding powder, in granulated form, is fed from a hopper into a heated cylinder where it is plasticised and forced into a die by hydraulic or screw pressure. The mould is prevented from opening by an external clamping force and the component remains there until set. The mould then opens and the component is ejected.

(a)

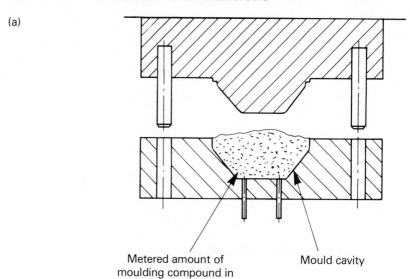

Metered amount of
moulding compound in
powder or pellet form

Mould cavity

((b)

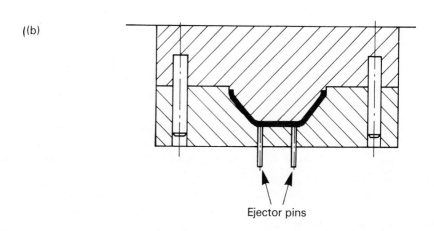

Ejector pins

Fig. 2.45 Compression moulding: (a) mould open; (b) mould closed, pressure
applied.

(a)

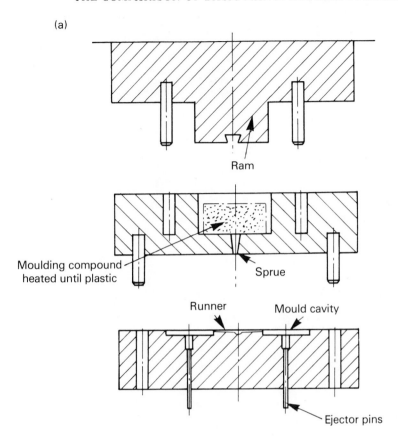

(b)

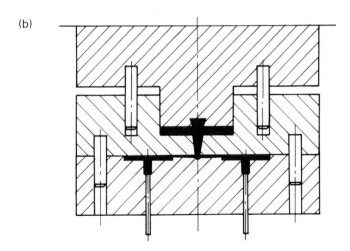

Fig. 2.46 Transfer moulding: (a) mould open;   (b) mould closed, pressure applied.

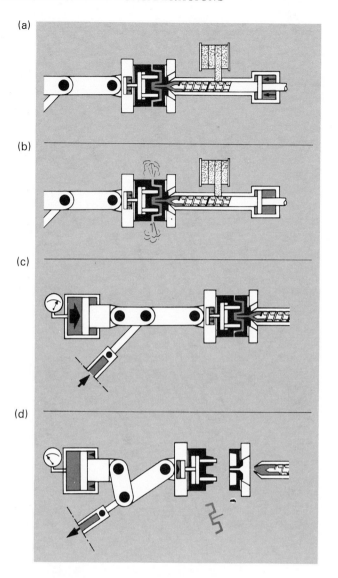

Fig. 2.47   Injection moulding:
   (a)   partial injection—injection of all or part of the material pre-heated by
         the screw after locking of the mould;
   (b)   degassing—relaxing of the locking cylinder to the chosen stroke. The
         gas is allowed to escape between the split line;
   (c)   end of injection-locking and if necessary final filling of the cavity;
   (d)   ejection—opening of the mould. The part is ejected. The screw pre-
         heats the material necessary for the following moulding.
   (By courtesy of Denis Leader (Machinery) Ltd.)

Injection moulding is a similar process to pressure die-casting for
metal and the basic machine and tooling costs are very high and it is
similarly used for large production quantities (see Fig. 2.47).

Table 2.1 *Components produced by injection moulding*

Thermoplastic material	Application
Nylon	Tennis rackets and strings, brush bristles, carburettor floats, tubing, handles
PVC	Bottles, buckets, bowls, window surrounds, traffic signs, electrical cable covering
Acetals	Motor car parts: heater fans, handles, gears and cams. Plumbing applications: taps, fittings, valves, glands
Acrylics	Telephones, sinks, baths, clock faces, goggles
Polythene	Consumer durables: bowls, buckets, brushes, waste bins, baskets, bottles
Polypropylene	Safety helmets, moulded chairs, crates, trim panels for motor cars
Polystyrene	Refrigerator parts, cups, shoe heels, cutlery handles, containers

Blow Moulding

This is a process used exclusively for thermoplastic materials. The material is extruded vertically downwards through a die in the form of a tube, called a 'parison'. A suitable length is clamped between two halves of a split die, sealing the tube ends, except where compressed air is introduced to blow the material against the walls of the die.

The process is used for the manufacture of bottles, containers and hollow toys (see Fig. 2.48).

Design Considerations When Plastic Moulding

The cost of producing dies for injection moulding is similar to that for pressure die-casting, and consequently this process is not adapted for short production runs. It is of paramount importance that the dies are correctly designed to achieve a satisfactory finished product, and often a single cavity mould may be built from epoxy resin for experimental purposes before proceeding with the construction of a multiple-cavity mould for large quantity production. An experimental moulding also allows the product designer to judge the suitability of the proposed design, as well as the mechanical and physical characteristics of the material of his choice, more critically than he could with a prototype which may have been machined from solid material.

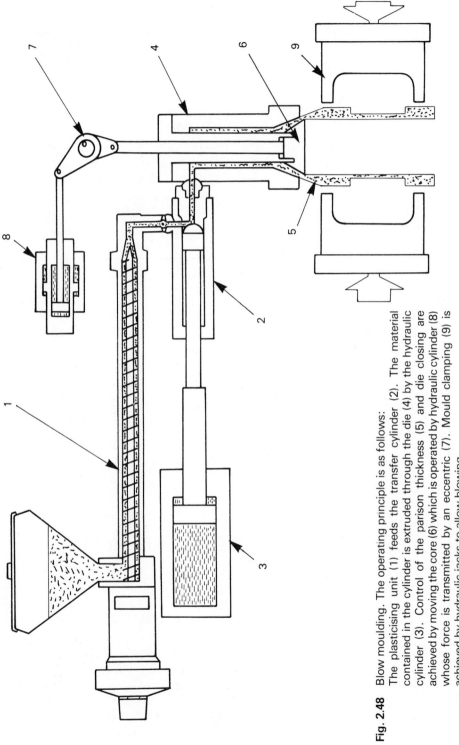

Fig. 2.48   Blow moulding. The operating principle is as follows:
The plasticising unit (1) feeds the transfer cylinder (2). The material contained in the cylinder is extruded through the die (4) by the hydraulic cylinder (3). Control of the parison thickness (5) and die closing are achieved by moving the core (6) which is operated by hydraulic cylinder (8) whose force is transmitted by an eccentric (7). Mould clamping (9) is achieved by hydraulic jacks to allow blowing.
(By courtesy of Denis Leader (Machinery) Ltd.)

(a)

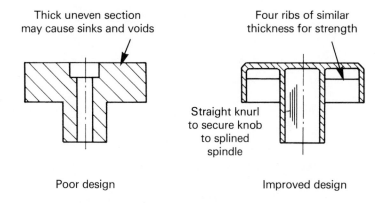

Thick uneven section
may cause sinks and voids

Four ribs of similar
thickness for strength

Straight knurl
to secure knob
to splined
spindle

Poor design

Improved design

(b)

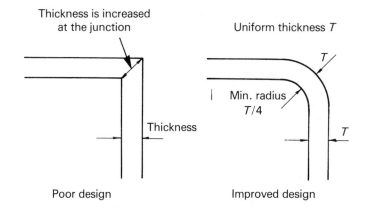

Thickness is increased
at the junction

Uniform thickness T

Thickness

Min. radius
$T/4$

T

T

Poor design

Improved design

(c)

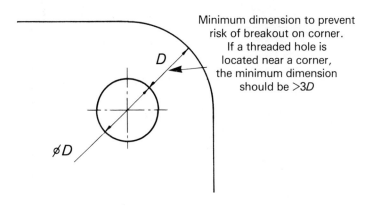

Minimum dimension to prevent
risk of breakout on corner.
If a threaded hole is
located near a corner,
the minimum dimension
should be $>3D$

D

$\varnothing D$

Fig. 2.49   Design aspects for plastic mouldings: (a) moulded knob;   (b) wall
thickness;   (c) location of cored holes.

Shrinkage occurs in a moulded component after it has been formed in the mould. The thinner sections cool more rapidly than the heavier and the thicker sections tend to shrink more than thinner sections causing warpage, sinks and voids. For this reason wall thicknesses should be kept as uniform as possible and, similarly, sharp corners should be avoided. Ribs may be provided to increase stiffness (see Fig. 2.49(a) and (b)). As in die-casting, all surfaces perpendicular to the parting line of the mould should be provided with a draft.

Cored holes should not be located too close to each other or too close to the edge of a component, thus helping to avoid a weakness in the design. This problem may be more acute when a threaded hole is to be cored (see Fig. 2.49(c)).

Inserts may be incorporated into mouldings, in a similar way to pressure die-castings, and may be secured by using machined flats or knurled portions (see Fig. 2.50). Screwdriver blades are secured into plastic handles by winging or flattening the blade (see Fig. 2.51).

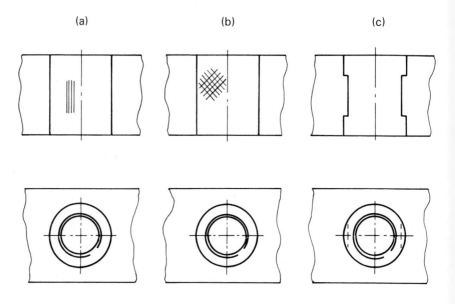

| (a) | (b) | (c) |

Fig. 2.50 Inserts 'moulded in' to plastic components:
  (a) straight knurled inserts preventing rotary motion;
  (b) diamond knurled inserts preventing rotary and axial motion;
  (c) two flats machined on inserts preventing heavier rotary and axial movement.

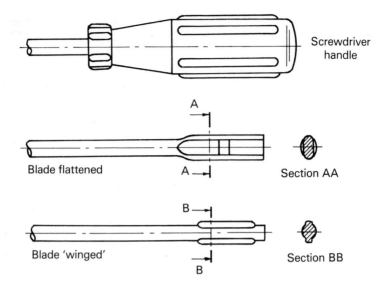

Screwdriver
handle

A

Blade flattened

Section AA

B

Blade 'winged'

Section BB

Fig. 2.51 Securing screwdriver blades in plastic handles.

Bosses are often used to reinforce a mounting position, particularly
in the corners of components. The height of the bosses should be
limited to twice the diameter (see Fig. 2.52). Care should be taken in
design to limit the introduction of a feather edge (see Fig. 2.53).
The design will be improved if a small inset is included which will
reduce the risk of break-out.

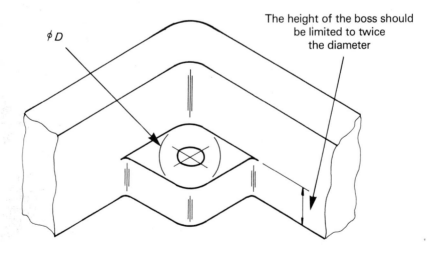

ϕD

The height of the boss should
be limited to twice
the diameter

Fig. 2.52 The use of the boss in plastic moulding.

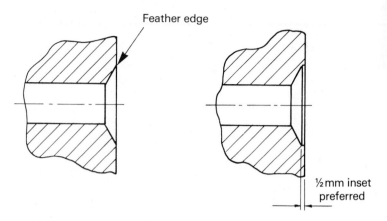

Fig. 2.53   The feather edge is limited in plastic moulding.

THE ADVANTAGES AND DISADVANTAGES OF EACH PROCESS

Advantages   ## Disadvantages

Bending by Hand and Folding Machining

* A wide variety of shapes can be bent and formed using simple low-cost equipment.

* Skilled labour is usually required.

* Depending upon the complexity of the bent article, the process may be time-consuming and expensive.

* The method is not suitable for large volume production.

* Only relatively thin material can be bent.

Press Brake

* A wide variety of thicknesses of material can be bent.

* A wide range of angular bends can be made using one set of tools, and the air-bending technique.

* Bends of considerable length can be made.

* The material is usually fed into press by hand.

* Hydraulic press brakes are slow in action and not suitable for volume production.

* The method is restricted to the bending of relatively simple shapes.

Advantages (cont.)	Disadvantages (cont.)

★ Mechanical-type press brakes, because of their rapid action, may be used for volume production of simple shapes using unskilled labour. They also may be adapted for simple piercing operations.

★ Hydraulic press brakes are capable of providing considerable bending pressures.

★ Low-cost tooling.

Press Tools

★ There is virtually no restriction on shape or size of component.

★ Very high production rates can be achieved.

★ Unskilled labour may be used.

★ The very high cost of tooling limits the use of press tools to the production of large numbers of components.

Hand and Open-tool Forging

★ A wide range of simple shapes and sizes can be produced in a variety of ferrous metals using simple tooling.

★ Strength and toughness are increased due to improved and orientated grain structure.

★ Forgings are free from blow-holes and porosity.

★ There is little material wastage.

★ Skilled labour is usually required.

★ The method is only suitable for low production runs.

★ The method is only suitable for components of relatively simple shape.

Drop Forging

★ There is very little limitation on the shapes that can be drop forged.

★ A wide range of ferrous and non-ferrous metals can be drop forged.

★ Strength and touchness are increased due to improved and orientated grain structure.

★ There is little material wastage.

★ A minimum of subsequent machining is required.

★ Semiskilled and unskilled labour can be used.

★ The high cost of dies restricts the process to high volume productions.

Advantages (cont.)   Disadvantages (cont.)

Extrusion

* A wide variety of shapes are obtainable, some of which cannot be achieved by any other means.
* Most metals with the exception of certain casting alloys can be extruded.
* Cold extrusion provides for a good finish and dimensional accuracy.
* Cold extrusions require little or no machining.
* There are improved mechanical properties due to grain flow.
* Interlocking sections can be obtained.

* The high cost of dies limits the process to large quantity production.

Gas Welding

* The low cost of the equipment means that gas welding can be undertaken at an economic cost.

* It is a labour-intensive process and is consequently only suitable for 'jobbing' work and thinner section materials.

Manual Metal Arc Welding

* The equipment is cheap.
* This is a versatile process easily set up.
* It can be used on most materials.

* It is a labour-intensive process.

Tungsten Inert Gas Welding

* This produces very high-quality welds, particularly good for thin metals.
* It can be done manually or automatically.
* The cost of equipment is low for manual application.
* It can be used for joining all metals except those with low boiling points like zinc.
* Because the electrode is non-consumable it does not add material to the weld pool.

* When material has to be added, i.e. in thick joints, it has to be added separately which can be a disadvantage.
* This method is not commercially competitive for welding thicker gauge metals.

Advantages (cont.)   Disadvantages (cont.)

Metal Inert Gas Welding

* This method has speed of versatility
* It can be operated automatically.
* It is particularly useful for welding aluminium.

* The special deoxidised wires are expensive.
* There is an inherent danger of cold lapping with some techniques.

Submerged Arc Welding

* Very high welding speeds are possible.
* Shielding of the operator is not required as the arc is buried under the flux.
* The process can be used outside in high winds which would disturb the gas shields in other processes.

* Flux and its handling equipment have to be used.
* Slag has to be removed.
* The process cannot be used on material thinner than 5 mm.
* Because of the application of granular flux this process has to be used in the horizontal position.

Resistance Welding: Spot Welding, Projection Welding and Seam Welding

* These methods can be robotised for production line welding.
* Miniature machines are available for the electronic industry.
* There is a consistent quality with automatic production.
* The high temperature of weld is localised.
* The weld is strong when subject to shear loads.

* The cost of robot machines is very high.
* Welds perform poorly in tension.

Resistance Welding: Resistance Butt Welding and Flash Butt Welding

* These methods are useful for joining two dissimilar metals where the amount of the expensive metal used can be minimised.

* These methods can only deal with a limited section size and area.

Advantages (cont.)

Friction Welding

★ This method is useful for joining two dissimilar metals where the amount of the expensive metal used can be minimised.

Electron-beam Welding

★ There is no atmospheric contamination.

★ Because of the vacuum it is possible to weld the more reactive metals successfully.

★ Only the minimum amount of material for fusion is melted, thus producing maximum joint strength with no distortion.

Sand-casting

★ This method is suitable for small quantity work. Although a pattern and a mould are required, they are relatively inexpensive when compared with moulds for pressure die-casting.

★ This is a flexible process. It can be used to produce castings from a few grammes to several tonnes.

Pressure Die-casting

★ Accurate non-ferrous components can be produced at a rapid production rate.

★ No machining of a die-cast component is required.

★ Surfaces are smooth and of pleasing appearance, requiring almost no mechanical preparation for painting or plating.

★ Inserts can be incorporated into a die-casting if the physical or mechanical properties need to be improved locally.

Disadvantages (cont.)

★ The method is limited to circular section bar or tubes.

★ Welds have to be formed in a vacuum.

★ Equipment is very costly.

★ Operators must be protected from radiation hazards.

★ The surface finish of casting is pitted and rough.

★ A new mould is required for each casting making the process labour-intensive.

★ There is a high capital cost of equipment and tooling.

★ Large production runs are required because of this high cost.

★ Only commercially viable for non-ferrous materials, particularly aluminium and zinc. Consequently this method is not suitable for highly stressed conditions.

Advantages (cont.)

Plastic Moulding

★ Plastics have excellent electrical insulation and chemical and corrosive resistance properties.

★ Self-coloured components are usually ready for use immediately after moulding.

★ Inserts can be incorporated into a plastic moulding for strength.

★ The finished product has a pleasing finish and 'feel'.

Disadvantages (cont.)

★ Plastics can only be used for low stressed conditions. They are not as strong as metals.

★ There is a high capital cost of equipment and tooling.

★ Large production runs are required because of this high cost.

★ The temperature resistance of plastics is low.

★ Plastics are brittle when cold.

★ There are environmental problems — it is difficult to dispose of scrap plastic components.

★ Plastics cannot be manufactured as precisely as metals.

★ Plastics absorb moisture and consequently show dimensional change.

A SUMMARY OF THE APPLICATIONS OF EACH PROCESS

Bending Using Hand Tools

This is a highly skilled process carried out by sheet metal workers and fabricators. The economics of this process restrict it to the production of very small numbers of sheet metal articles.

Bending Using the Press Brake

A wide range of material sizes and thicknesses can be bent using this machine provided that the components are not too complex in shape. Different angular bends may be made using a single set of tools and the air-bending technique. Variable-speed mechanical press brakes may be used for small or large quantity production.

Bending Using Press Tools

Bending is only one operation that may be carried out using press tools, others being piercing, blanking and forming. The high cost of press tools limits their use to the mass production of sheet metal parts in a variety of shapes and sizes.

Hand Forging

This process is generally restricted to the production of very small numbers of simply shaped components in plain carbon or alloy steel.

Open-tool Forging

Forgings from a few kilograms to several tonnes can be produced in plain carbon and alloy steels. Some restriction on shape is imposed due to the relatively simple tooling used. Open-tool forging is generally used for small quantity or batch production.

Drop Forging

A wide variety of both ferrous and non-ferrous metals can be drop forged, but owing to the high cost of the dies this process is restricted to the production of large quantities of components. Drop forging brings the component very close to its finished shape and size, machining is therefore kept to a minimum. As with all forging processes, component strength is improved due to the grain flow following the shape of the component.

Extrusion

An almost unlimited variety of cross-sectional shapes can be produced in various ferrous and non-ferrous metals and alloys. Extrusions are suitable for making components that have the same cross-section throughout their length.

Extruded section is suitable for both large and small quantity production as a wide range of standard sections are available.

Gas Welding

The most widely used gas welding process is oxyacetylene welding. It is very economical and because of low-cost equipment can be used efficiently for small quantity work.

Arc Welding

Arc welding covers a range of processes using manual and automatic production methods. The welds produced are superior to those made by the gas welding process.

Resistance Welding

The cost of equipment for this process ranges from moderate to high. It can be suitable for 'batch-work' or high volume production work depending upon the process chosen and the amount of automatic handling to be introduced.

Friction Welding

This method is only suitable for the welding of basically round-section components. The cost of a friction welding machine is quite high, but the cost of the process allows components to be welded in batch quantities economically.

Electron-beam Welding

Very high-quality welds are produced by this process but the equipment is very expensive. Consequently used for high volume production or where high-quality welds are a necessity, i.e. space vehicles.

Sand-casting

Sand-casting is used for the manufacture of small to medium-sized metal objects — both ferrous and non-ferrous — in quantities ranging from one-offs to small batches. Surface finish is quite rough and fine detail cannot be coped.

Pressure Die-casting

Pressure die-casting is used to make large quantities of small non-ferrous metal components, mainly aluminium and zinc alloys. The surface finish and accuracy obtained are excellent. A subsequent surface coating can be given — from painting to gold plating — with hardly any mechanical preparation work. The high cost of dies prohibits the manufacture of small quantities.

Compression and Transfer Moulding

Both these processes are used for the manufacture of thermosetting plastic components. Moulds are expensive to produce, and consequently these processes are not suitable for small quantity production.

Injection Moulding

This method of moulding is one with the widest application. All thermoplastic materials can be moulded by this process, and the moulding produced has an attractive appearance and is ready for use without any 'post-forming' operations being necessary. This is a similar process to die-casting for non-ferrous metals. The high cost of dies prohibits the manufacture of small quantities.

Blow Moulding

The blow moulding process is used to make hollow components in thermoplastic materials. It is ideally suited for the manufacture of containers up to 1000 litres capacity in medium to large quantities.

CHOOSING A SHAPING PROCESS ———————

This chapter has covered some aspects of the six common approaches for shaping materials. Some applications of the processes overlap in capability, and a choice may have to be made between equally suitable processes. This choice will be influenced by the type of material to be used.

Let us now consider a manufacturing problem: *The Production of a Body for a Handyman's Knife* similar to that shown in Fig. 2.54. A typical specification would involve a number of dimensions including some within the tolerance of 0.1mm. The body is made in two halves which are fastened together, on assembly, with a screw.

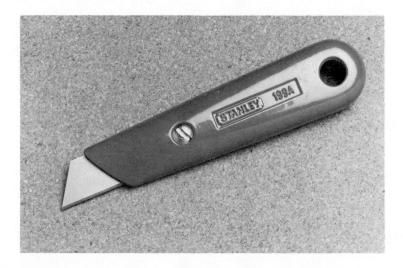

Fig. 2.54   A handyman's knife.   *(By courtesy of Stanley Tools Ltd.)*

The possible processes can now be considered in turn:

(i) *Machining.* Basically any component can be produced by machining. In this instance however it would be a very wasteful process, both in material costs — a large amount of its volume would be machined away — and in labour costs — in view of the length of time needed to machine a shape of this contour. This method of production is discounted.

(ii) *Sand-casting.* This process could be used to produce the two halves of the body. The castings would need some machining to produce the threaded hole, the rack locating the blade shifter and certain precise dimensions for location when the two halves are fastened together.

Wooden patterns would be required but no expensive equipment is necessary. Material costs are low as little scrap would be produced. Materials most suitable for the process and component are aluminium, zinc or bronze alloys. The component would need to be painted.

(iii) *Pressure die-casting.* The tolerance on dimensions of components made by this process are closer than those produced by sand-casting. This would eliminate all the machining, with the possible exception of the tapped hole. Expensive dies would have to be made, but the surface finish of the die-casting would need very little mechanical preparation prior to painting. Ideal materials for this process and component would be aluminium or zinc alloy.

(iv) *Injection moulding.* This process would need the body halves to be made from a thermoplastic material. The mouldings would be self-coloured with a very good surface finish and painting would not be necessary. A screwed metal insert could be moulded in to one half of the body to accept the securing screw, after assembly. Expensive dies would have to be made.

(v) *Bending and welding.* The body halves could be produced by bending, forming and piercing. This would necessitate additional machining work — to produce the tapped hole and the shifter rack. The latter item may have to be welded or soldered into position. Materials suitable for this process and component include mild steel and brass sheet. The component would need to be painted.

(vi) *Extrusion.* This process could not be used in the manufacture of the body halves. Extrusion requires the component to have the same cross-section along its whole length.

(vii) *Forging.* The two body halves could be produced as forgings, although a lot of work would be needed to improve the quality of the surface after forging. Some machining would be necessary similar to that undertaken if produced as sand-castings. The component would need to be painted.

We have briefly examined several different processes for producing the body for the handyman's knife. Two processes, machining and extrusion, have already been eliminated as unsuitable. However, one of the factors which will be crucial in helping us to arrive at a sound decision is the *production quantity*.

A product of this type will command a fairly low selling price and to justify its investment a company would have to manufacture large quantities. Modestly, sales quantities of 20 000 per annum could be envisaged with a minimum sales life of 15 years.

This now helps to put the choice of process into a clearer perspective for we can immediately eliminate some of the other processes such as:

(i) *Sand-casting.* This is a process suitable for small/medium batches and a new mould would be needed for each casting. The surface finish is poor and additional work would be necessary on each and every casting to resolve this.

(ii) *Bending and welding.* The basic shape could be produced, economically, in large quantities by presswork, but the cost of machining and securing a shifter rack would be too high.

(iii) *Forging.* Too much post-forging work would be necessary to improve the surface finish. Additional expensive machining would have to be undertaken.

The final choice lies between die-casting and injection moulding. In each case expensive dies have to be made, but the cost will be justified by the production quantity.

The total cost of production, as an injection moulding, may be 10–15% cheaper than a die-casting, but it has been decided in this case to use a zinc pressure die-casting. The hand-tool trade is a traditional market, and a painted metal body has often more appeal than a lighter plastic body. This is a policy decision based on market factors rather than the logic of production economics.

QUESTIONS

1. Which of the processes examined in this chapter would be used for the manufacture of the following products:

(i)   a motor-car bonnet;

(ii) a plastic bottle;

(iii) an engine connecting rod;

(iv) a screwdriver handle;

(v)   a washing-machine agitator;

(vi) a frame for a free-standing drawing board;

(vii) a central heating radiator;

(viii) a body for a 13 amp plug;

(ix) the pins for a 13 amp plug;

(x)   a sliding door track;

(xi) an eyebolt used in a lifting device;

(xii) a carburettor body;

(xiii) a ring spanner;

(xiv) a fountain pen nib?

2. In order to reduce the amount of high-speed steel used when producing high-speed steel drills $> \emptyset\, 20\, mm$, the shank is manufactured
from low carbon steel and welded to the body. State and briefly
describe a suitable welding process for this type of joint.

3. Describe briefly a welding process suitable for the joining of motor-
car panels.

4. Why does arc welding produce a better joint that gas welding?

5. Briefly describe the pressure die-casting and injection moulding
processes and compare their differences.

6. Comment upon the major differences between sand-casting and
pressure die-casting.

7. Describe two ways in which bends can be produced by a press brake.

8. State the reason why extruded section is suitable for small or large
quantity production.

9. A tow bar assembly comprises a welded steel bracket and a drop
forged towing ball. Give four reasons why you think drop forging has
been chosen to make the towing ball.

10. Using sketches list two important component design considerations
for each of the following manufacturing processes:

(i)   sand-casting;

(ii) plastic moulding;

(iii) open-tool forging;

(iv) press tools.

ASSIGNMENT ————————————————————————

Critically examine various processes suitable for the manufacture of 1000 microscope bodies. Comment upon their suitability/unsuitability for this product and state a concluding process giving the reasons for your choice.

DESIGN

Ergonomic Factors of Design

After reading this chapter you should be able to:

★ analyse the ergonomic factors involved in design (G);

★ define and state the meaning of ergonomics in terms of man-machine relationships (S);

★ describe the ergonomics control loop as display element, communication channel, decision-making channel, the control element and the control (S);

★ analyse the arrangement of dials on a control panel for sequence and compatibility (S);

★ analyse the ergonomic factors involved in the use of simple instruments (S);

★ analyse the ergonomic factors involved in the use of simple tools (S).

(G) = general TEC objective
(S) = specific TEC objective

ERGONOMICS

Ergonomics is the scientific study of the relationship between man and his working environment. The environment will include not only the conditions in which he is expected to work but also the materials and tools with which he works. The industrial demands of today require that man and his machine (and in this sense his machine may be a lathe, a motor car, an instrument or a hand tool) need to be an efficient production unit.

In order to maintain productivity it is necessary to consider the environmental effect on work performance and study the principles of movement and the layout of work places to improve efficiency. These aspects of ergonomics are covered in another TEC Unit, Control of Manufacture III.

The designer needs to design a machine in order to help the operator and make his job reasonable and possible to undertake, so that it will reduce physical and mental strain and allow him to devote his attention to the judgement needed to undertake the job. If the machine is to be effective, the operator must be able to control it easily and with accuracy. The amount of force needed to operate it must be within the easily repeatable capacity of the operator. The positioning of the control knobs, switches or levers must be such that they can be used effectively.

The position and types of instruments on a machine must be considered. Instrument panels should be designed so that information

133

can be displayed to show the person just what he needs to know. It is not good practice for a dial instrument to show additional information which is unnecessary. The average car driver does not need to know the precise water temperature in his engine. It is only necessary to know whether the temperature is within an acceptable range and a well-designed dial will give this information at a glance.

Instrument dials give information to the operator, and it is often necessary for him to make some correction, such as turning a control knob or flicking a switch or operating a lever. Controls should be designed in order that they may be operated by the large majority of users. If the controls are to be operated by men and women, consideration must be given to this at an early design stage. The directions of movement must have compatibility – which is a topic discussed in greater detail later in the chapter. Switches need to be moved down for 'on' and up for 'off'. Increases in force, time or temperature should be attained by movement to the right or upwards, decreases to the left or downwards. Variations of this may occur however where an operator is making specific movements with each hand (see p. 136).

The man-machine relationship will be more efficient if attention is given to the quality of the environment. Consideration must be given to the provision of light and heat, the control of noise levels, and the general brightness and colour of the working area. All operators feel an uplift when the workshop is newly painted with light-coloured walls reflecting the light and brightening what may have been a dismal and cheerless environment.

Colour may also be used to give identification or coding for safety purposes, making the operator's job safer. Black and yellow diagonal banding is used to denote dangerous hazards. The inside of machine guards are sometimes painted yellow to warn of danger when open. Electrical equipment is painted orange and can be located more easily in consequence.

The positioning of machines in a production unit will help man/machine relationships, for it will give easier access and transportability as well as improvement in work flow.

Thought must be given to the operator's posture and comfort. The best results will be obtained when he or she is operating in a less tensed position. The driver of a motor car should be sitting in a relaxed attitude with good wide vision and adequate ventilation to keep him fresh and alert for his task.

Many of these considerations may be regarded by the designer as just 'commonsense', but there are still many examples in industry and the home where it appears to have been ignored. Cars are still being marketed where the bonnet release lever is located by the passenger's left foot!

THE ERGONOMICS CONTROL LOOP

The ergonomics control loop is the relationship between the operator and the machine he controls and is shown in diagrammatic form in Fig. 3.1.

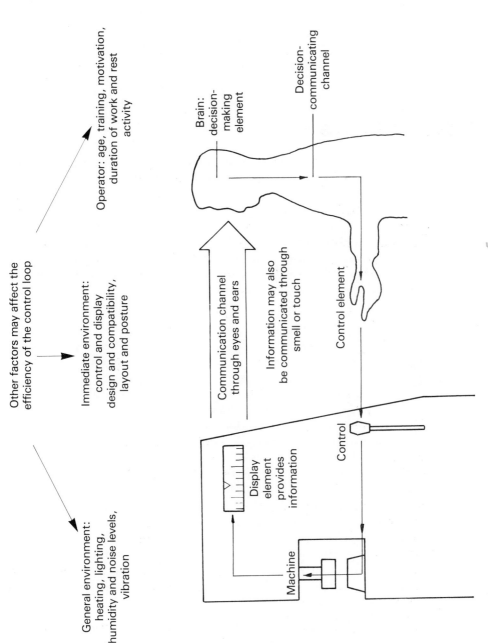

Fig. 3.1 Ergonomics control loop.

The *display element* provides information of how the machine is performing the operation. The *communication channel* sends this information to the brain of the operator generally through his eyes and ears but sometimes through one of his other senses. The *decision-making element* is the brain of the operator, which feeds the message through his central nervous system, the *decision communication channel*, to his hand, the *control element*, which can adjust the *control* and reset the machine to complete the control loop. Other factors, such as the immediate and general environment and the age, training, motivation and fatigue of the operator, may have an affect on the efficiency of the control loop.

CONTROL PANELS

The main function of an instrument is to supply information to the user. The designer, who is responsible for the instrument layout, has to make sure this is achieved quickly and accurately. If the display is badly designed, the information will still be communicated to the user but may add to his physical and mental strain which could promote error of action. Communication is a two-way function, and the user may need to take action to respond to the information and consequently the designer needs to aim for a successful working unit — man and machine.

The directions of movement must have compatibility — they must be consistent. Electric light switches need to be moved down for 'on' so this has become a learnt pattern of behaviour. Wherever a person is in the UK he would expect a light switch to operate in this way. Similarly clockwise-to-increase is a learned pattern. The expected relationships of movement between controls and displays are shown in Fig. 3.2. If this system is adopted during the design process, the learning time for the operation of the equipment will be shorter and there will be less risk of accident, since in an emergency the operator will revert to the expected direction of movement. Operator fatigue and strain will also be reduced. Variations of this may occur, however, where an operator has to make switch-turning movements from a fixed seated position, e.g. in a motor car. In such cases increases are more logically made by turning either wrist in an outer movement. Thus increases with the right hand would be made clockwise and increases with the left hand would be made anticlockwise. Where controls are connected with dials in a panel, they should both be designed with the same spatial relationship: the right-hand dial of a row should be operated by the right-hand control, etc.

Examples of instrument and control compatibility in a motor car facia panel are shown in Figs. 3.3 and 3.4. The general view of a motor car facia panel is shown in Fig. 3.5.

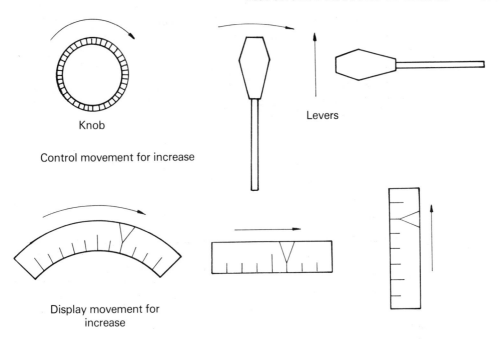

Fig. 3.2 Relationships of movement between controls and displays.

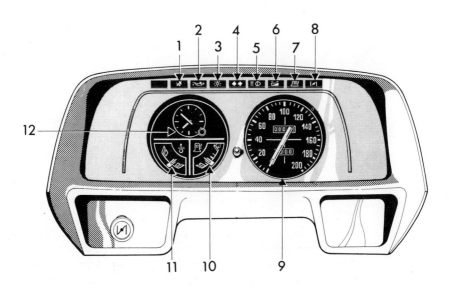

Fig. 3.3 A motor car instrument panel. 1-8 driver warning lights. Dial movements: 9, the speedometer needle (km/h) rotates clockwise with increasing speed; 10, the fuel gauge needle moves up as tank is filled; 11, the needle moves upwards as the temperature increases; 12, the hands move clockwise as time proceeds. *(By courtesy of the Talbot Motor Co. Ltd.)*

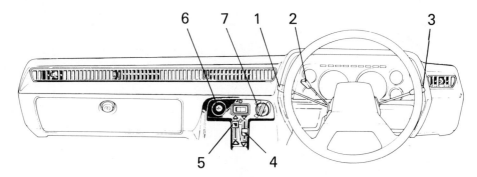

Fig. 3.4 A motor car control panel. Manual movements: 1, the left hand moves the switch anticlockwise for side and rear lamps and headlamps, up for headlamp dip and pulls the switch for headlamp flash; 2, the left hand pushes the lever up for right turn and down for a left turn; 3, the right hand pushes the lever down for 'on' and pulls it for a screen wash; 4 and 5, the left hand pushes the levers up for increasing air flow to the interior and screen respectively; 6, the left hand pushes in to operate the cigarette lighter; 7, the left hand turns the knob anticlockwise to increase the temperature.   *(By courtesy of the Talbot Motor Co.Ltd.)*

Fig. 3.5 A general view of a motor car facia panel. *(Photo: John Wright. Reproduced by permission of the Talbot Motor Co. Ltd.)*

If we consider the layout of a motor car facia panel, two principles could be applied to the design:

(i) the operation may follow a fixed sequence, or

(ii) the controls and displays may be grouped according to their functions.

Included in the first category should be the 'engine starting and getting ready to move' sequence:

choke;
ignition;
ignition and oil warning lights;
starter;
handbrake warning light;
gear lever;
handbrake.

Some of the controls may be used simultaneously, and as the right hand is the preferred hand the designer has to decide which operation should be undertaken with this hand.

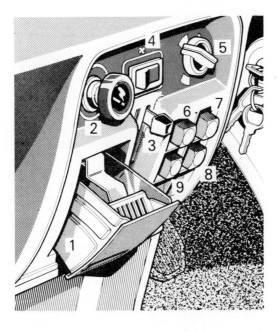

1. Ash tray
2. Cigarette lighter
3. Air control lever
4. Blower switch
5. Heater temperature control
6. Switch: hazard warning
7. Warning light and test switch for front brake pads
8. Switch: rear window heating
9. Switch: rear fog lamps

Fig. 3.6 The functional grouping in a motor car facia panel. These controls are situated to the left of the steering column and consequently are operated by the left hand. *(By courtesy of the Talbot Motor Co. Ltd.)*

1. Forward and reverse screw cutting and feed selector
2. Low and high feed and screw cutting range selector
3. Oil level
4. Spindle speed selectors
5. Four speed and screw cutting selectors
6. Coolant 'off/on'
7. Power 'on' light
8. Motor 'stop/start'

Fig. 3.7   The functional grouping of controls on the headstock of a centre lathe (Colchester Triumph 2000). *(By courtesy of The Colchester Lathe Co. Ltd.)*

In the latter category the controls and displays should be grouped relative to their importance and more frequent use. Consideration may be given to:

    (i) lever switches for side and rear lamps, headlamp dip, headlamp main beam and headlamp flash together with their appropriate warning lights;

    (ii) heater temperature control, blower switch and heater control levers for car interior and windscreen;

    (iii) windscreen wiper and washer levers;

    (iv) cigarette-lighter and ash-tray.

Figs. 3.6 and 3.7 show examples of functional grouping.

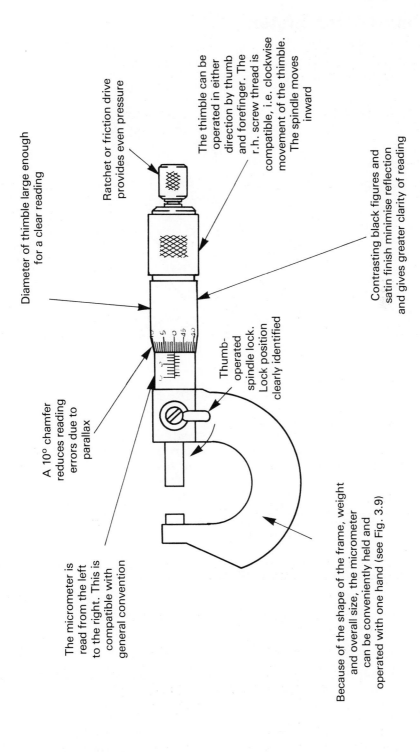

Diameter of thimble large enough for a clear reading

Ratchet or friction drive provides even pressure

The thimble can be operated in either direction by thumb and forefinger. The r.h. screw thread is compatible, i.e. clockwise movement of the thimble. The spindle moves inward

Contrasting black figures and satin finish minimise reflection and gives greater clarity of reading

A 10° chamfer reduces reading errors due to parallax

Thumb-operated spindle lock. Lock position clearly identified

The micrometer is read from the left to the right. This is compatible with general convention

Because of the shape of the frame, weight and overall size, the micrometer can be conveniently held and operated with one hand (see Fig. 3.9)

Fig. 3.8   The external 0 – 25 metric micrometer.

INSTRUMENTS

The Metric Micrometer

Fig. 3.8 examines the ergonomic factors related to the 0-25 metric micrometer. It is interesting to note that the micrometer is designed to be held and read simultaneously only by a right-handed person (Fig. 3.9).

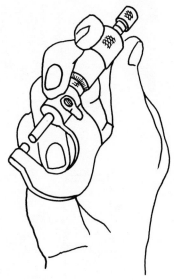

Fig. 3.9   The method of holding the 0 – 25 metric micrometer.

Reading the micrometer is compatible with general convention in that it is read from the left to the right, a well-formed habit. The reader might compare reading the scales of a 0-25 external micrometer with those of a depth micrometer which is read from right to left.

Information regarding the design of external micrometers is given in *British Standard* 870:1950.

The Mechanical Comparator

A comparator is an instrument which, after being set from a standard (i.e. slip gauges), compares the length introduced with that to which the instrument has been set.

Fig. 3.10 analyses the ergonomic factors of a mechanical comparator.

It will be seen that certain controls, because of their position, should be operated by the right hand and others by the left hand.

The method of indication is clear and of suitable magnification, e.g. 300-5000 times. The comparator has a 'dead beat' mechanism which ensures that the indicating pointer is free from oscillation.

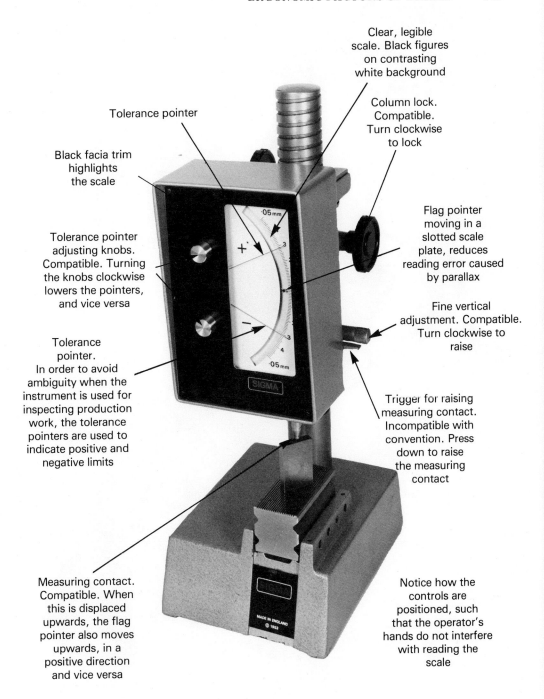

Clear, legible scale. Black figures on contrasting white background

Column lock. Compatible. Turn clockwise to lock

Tolerance pointer

Black facia trim highlights the scale

Flag pointer moving in a slotted scale plate, reduces reading error caused by parallax

Tolerance pointer adjusting knobs. Compatible. Turning the knobs clockwise lowers the pointers, and vice versa

Fine vertical adjustment. Compatible. Turn clockwise to raise

Tolerance pointer. In order to avoid ambiguity when the instrument is used for inspecting production work, the tolerance pointers are used to indicate positive and negative limits

Trigger for raising measuring contact. Incompatible with convention. Press down to raise the measuring contact

Measuring contact. Compatible. When this is displaced upwards, the flag pointer also moves upwards, in a positive direction and vice versa

Notice how the controls are positioned, such that the operator's hands do not interfere with reading the scale

Fig. 3.10 The mechanical comparator. *(Photograph by courtesy of Sigma Ltd.)*

The shaped handle enables the hand to keep a natural fatigue-free position and firm grip to provide the cutting force and maintain alignment.

Cutter alignment lever

Knob to adjust depth of cut. This can be operated by the thumb and forefinger from the pistol grip position. Note compatibility, as the knob to adjust depth of cut is turned clockwise, the depth of cut is increased.

Cam clamp-shaped to accommodate thumb and forefinger

The ball-shaped knob fits in the palm of the hand and allows the hand to take up a natural position to provide the downward pressure and maintain the depth of cut

Fig. 3.11   The bench plane.   (*Photograph by courtesy of Stanley Tools Ltd.*)

TOOLS ───────────────────────────────

The Bench Plane

Consider the bench plane shown in Fig. 3.11. It is designed such that it can be held comfortably in both hands by either a right- or left-handed person, without the hands having to be contorted into unreasonable positions.

The sculptured handles allow the hands, wrists and arms to take up the most natural and optimum position for (a) providing the most effective cutting force and (b) maintaining alignment of the plane during cutting, with the minimum of user fatigue.

Controls for adjustment of the cutting blade are designed so that they are easily operated without the user having to remove both hands from the plane.

An important aspect of ergonomics is safety. The user has little chance of cutting himself on the blade unless he runs his hands over the underside of the plane. The body, handles and ancillary details are radiused or chamfered and of suitable finish such that the user has little chance of sustaining an injury provided the plane is used correctly.

The Engineer's Steel Tape Measure

Fig. 3.12 shows an engineer's steel tape measure that has been carefully designed, with ergonomic considerations in mind.

Notice how the tape housing is designed such that its knurled contours fit the palm of the hand, enabling it to be held and operated in a comfortable and secure manner by either hand.

The tape may be locked in any desired position by moving the knurled slider, with the thumb in a downwards motion. Being able to lock the tape assists in making the reading of the tape much easier. When the slider is moved upwards the tape automatically retracts into its housing.

To enable rapid and precise reading the tape has a non-reflecting yellow finish on which contrasting black, metric and imperial scale divisions are marked. These scale divisions being no smaller than 1 mm and $\frac{1}{16}$ inch respectively.

The length of the tape housing is indicated in centimetres, and in inches on the reverse side. When using the tape on an inside surface this length can be added to the actual reading taken from the tape.

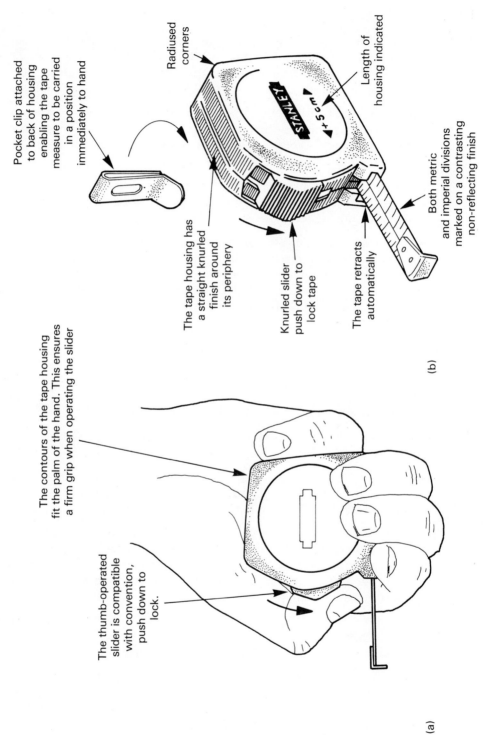

Pocket clip attached to back of housing enabling the tape measure to be carried in a position immediately to hand

Radiused corners

Length of housing indicated

The tape housing has a straight knurled finish around its periphery

Both metric and imperial divisions marked on a contrasting non-reflecting finish

Knurled slider push down to lock tape

The tape retracts automatically

(b)

The contours of the tape housing fit the palm of the hand. This ensures a firm grip when operating the slider

The thumb-operated slider is compatible with convention, push down to lock.

(a)

Fig. 3.12   The engineer's steel tape measure.   *(By courtesy of Stanley Tools Ltd.)*

QUESTIONS

1. Describe the ergonomics control loop, stating each element and showing how an operator uses it to control the machine.

2. State some of the factors which affect the efficiency of the ergonomics control loop.

3. Examine a motor car facia panel and analyse the arrangement of dials for sequence and compatibility.

4. Using simple line diagrams analyse the ergonomic factors related to the following pieces of equipment:
   - (i) a 'mole' self-grip wrench;
   - (ii) a screwdriver;
   - (iii) an adjustable spanner;
   - (iv) a pair of combination pliers;
   - (v) a hacksaw;
   - (vi) a vernier caliper;
   - (vii) a depth micrometer;
   - (viii) a handyman's knife.

5. Explain the importance of compatibility in the case of:
   - (i) the rotation of a car's steering wheel and the movement of its front wheels;
   - (ii) the rotation of a screw and the direction of its travel;
   - (iii) the movement of the control column of an aircraft and the changes in direction of flight.

6. What are the effects of the following to man/machine relationships:
   - (i) layout of machines in a workshop;
   - (ii) colour, temperature and light of the working environment;
   - (iii) design of instruments and layouts of controls.

7. Colour coding is often used to identify or to show a potential hazard. What are the dangers in over-applying colour and making the workshop look like a fairground?

Safety Aspects of Designs

4

After reading this chapter you should be able to:

★ consider the safety aspects of designs (G);

★ use the safety regulations, statutory and published for recommended safety practice (S);

★ evaluate the balance between safety and design requirements for products (S);

★ state the ergonomic aspects of safety and the use of safety devices such as guards and covers (S);

★ state the use of interlock arrangements, safety valves, fuses, governors, warning lights and buzzers (S);

★ state the concepts of the fail-safe system and discuss fail-safe devices (S).

(G) = general TEC objective
(S) = specific TEC objective

SAFETY REGULATIONS

A major factor in the promotion of safety standards is the legislation 'Health and Safety at Work, etc. Act, 1974'. The purpose of the act is to provide the legislative framework to promote, stimulate and encourage high standards of health and safety at work.

The act is an enabling measure superimposed over existing health and safety legislation contained in 31 relevant acts. These range from the Explosive Act, 1875, the Factories Act 1961, the Offices, Shops and Railway Premises Act, 1963 to the Employment Medical Advisory Service Act, 1972. The new Act provides for the gradual replacement of existing health and safety requirements by revised and updated provisions, in the form of a system of regulations and approved codes of practice prepared in consultation with industry. The objective of the new provisions is not only to rationalise and up-date the law but to improve the standards of protection which it affords to people at work and the public.

The act covers all 'persons at work' whether employees, employers, or self-employed, with the exception of domestic servants in a private household. About five million people, such as those employed in education, medicine, leisure industries and in some parts of the transport industry who have not been covered by previous health and safety legislation, are now protected for the first time. The legislation protects not only people at work, but also the health and safety of the general public who may be affected by work activities.

The designer now has a major obligation in the safety aspects of designs for it is necessary, under the terms of the act, for him to design products which are safe in use and not injurious to health. Part I, paragraphs 6(1) of the act states:

> 'It shall be the duty of any person who *designs*, manufactures, imports or supplies any article for use at work:
>
> (a) to ensure, so far as is reasonably practicable, that the article is so designed and constructed as to be safe and without risks to health when properly used;
>
> (b) to carry out or arrange for the carrying out of such testing and examination as may be necessary for the performance of the duty imposed on him by the preceeding paragraph;
>
> (c) to take such steps as are necessary to secure that there will be available in connection with the use of the article at work adequate information about the use for which it is designed and has been tested, and about any conditions necessary to ensure that, when put to that use, it will be safe and without risks to health.'

It is intended that the basic obligation should, at each stage of production and supply, lie on those who have primary control. Health and safety considerations must be taken into account in the design and making of such products. This is without prejudice to the basic obligation on employers to ensure safety in the provision and use of machines and materials. If a manufacturer receives a written acceptance of the safety responsibility, then the accountability is transferred to the employer. For example, a shipbuilder can accept the responsibility of testing lengths of cable when it has been supplied by a manufacturer. The questions a designer must ask himself are detailed in Fig. 4.1.

Statutory Duties

Parliament has from time to time legislated upon matters relating to safety and health with a series of statutes, such as the Factories Act, 1961. In these acts parliament has tried to place a minimum standard of care upon employers to provide and maintain safe working conditions, the aim being preventive rather than punishment by imposition of criminal penalties.

The Health and Safety at Work, etc. Act, 1974 replaces those separate acts by a single comprehensive act and it intends to give a minimum standard of care by those people considered in the act. It is 'penal' in its authority — i.e. with the possibility of imprisonment.

The statutory duties of employers are that they must safeguard as far as is reasonably practicable, the health, safety and welfare of the people who work for them. This applies in particular to the provision

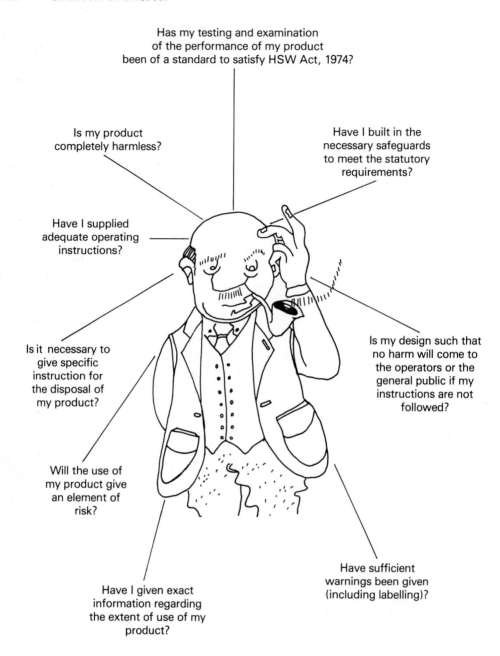

Fig. 4.1 The questions a designer must ask himself when considering the safety aspects of designs.

and maintenance of safe plant and system of work, and covers all machinery, equipment and appliances used.

The act requires that certain classes of plant and equipment must receive periodic attention. Some of these are listed below in Table 4.1.

Table 4.1 *The legal requirements for examining equipment*

Equipment	Statutory interval of attention	Schedule of requirements
Hoists and lifts	6 months	Every hoist or lift shall be thoroughly examined by a competent person at least once in every period of 6 months.
Chains, ropes and lifting tackle	6 months	These shall be thoroughly examined by a competent person at least once in every period of 6 months.
Cranes and other lifting machines	14 months	All such parts and gear shall be examined by a competent person at least once in every period of 14 months.
Steam boilers	14 months	All steam boilers shall be examined at least once in every period of 14 months.
Air receivers	26 months	Every air receiver shall be thoroughly cleaned and examined at least once in every period of 26 months.
Fire warnings	3 months	There shall be tested, at least once in every period of 3 months, every means of giving warning in case of fire.

The student should now examine some of the books listed under *Further Reading* bibliography at the end of the book to enhance his understanding of safety requirements.

THE BALANCE BETWEEN SAFETY AND DESIGN —

When a product is being designed, safety must be a primary consideration. An article cannot be marketed which has a fair element of risk attached to its use. The user must be confident in the working facility of the product and have confidence in its ability to perform its particular task with no danger to himself or any member of the public.

A balance must be achieved, however, between the safety and the function of the product, for while safety must be the foremost thought in the designer's mind it is not the only factor. If the designer concentrates all his efforts into the safety aspect he may not satis-factorily design for function, and in consequence may make the product very difficult, if not impossible, to use for the duty intended. This may then introduce another element of risk to the person using the product.

Safety needs to be promoted at the design stage of a product. It does not want to appear that the article was designed and then the safety aspects considered as an afterthought, for they must be a major feature of the product.

The balance between safety and design requirements for a machine tool is illustrated in Fig. 4.2.

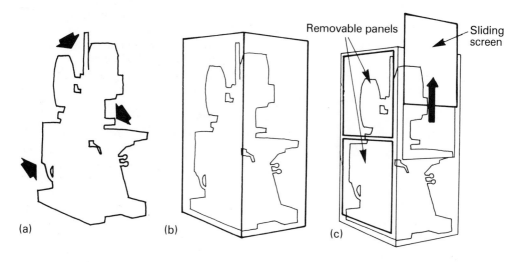

Fig. 4.2 The balance between safety and design requirements:
(a) a hypothetical machine in which, by virtue of its design and usage, several hazards are present;
(b) the same machine safeguarded completely from the hazards within but lacking the access that working the machine obviously requires;
(c) a method of providing front access for operation and removable side panels for maintenance or inspection purposes.

Fig. 4.2(a) represents a hypothetical machine which has been designed to undertake a particular function. In the course of usage, however, a number of hazards are present which need to be safeguarded from the operator.

Fig. 4.2(b) represents the same machine which has been safeguarded completely from the hazards within but lacking the access that a working machine obviously requires. Consequently the machine cannot perform the function for which it was intended.

Fig. 4.2(c) represents the same machine which has a method of providing front access, by means of a vertically sliding screen, which cannot be opened whilst the dangerous parts are in motion. This allows the machine to perform its task and at the same time affords the safety protection which is necessary for the operator. Removable panels are provided at the side of the machine for maintenance or inspection.

THE ERGONOMICS OF SAFETY ─────────────────────

The ergonomic factors involved in design have already been examined in Chapter 3. Let us now consider the ergonomic aspects of safety.

It has been previously stated in Chapter 3 that for a man to work to his best effect he needs to be physically and mentally at ease. The chances of mistakes being made are reduced when a man is not working under stress, and, as mistakes can lead to accidents, consideration of ergonomic factors will promote greater efficiency by establishing good and safe working practices.

The effect of errors can be lessened by applying the principles of physiology, psychology and anatomy to the relationship between a man, his machine and his working environment.

Physiology deals with the work function and the surrounding conditions in which the operator works and their effect upon him. Some of these effects were considered in the previous chapter and are listed in Fig. 3.1. They include the general level of heat, light, noise and humidity and the design, display, compatibility and layout of controls. All efforts must be made to ensure that displays are clear and positive. Colour can provide a most useful contribution to visual display. Vision, however, is a sense which is most easily overloaded. Just consider driving down a busy thoroughfare in any of our major cities. The number of road warning signs which are in close proximity to each other and which a driver must observe may be so numerous that he could be forgiven for failing to take notice of a particular sign. Sometimes it is necessary to reinforce the visual display by another means which will attract an operator through another of his senses.

Psychology is concerned with the aspects of industrial life and safe working. People are creatures of habit. They react automatically to certain signals and expect certain things to result from certain actions. For this reason it is highly desirable to standardise displays and controls on machines and equipment, so that a person can take the necessary action automatically to give the required adjustment.

The motor industry could standardise the elements of car controls more satisfactorily than is presently apparent. The handbrake lever is placed either to the left or right of the driver or by the side of the steering column. If a driver who regularly drives a certain type of car suddenly changes to drive another type he may find he releases his hand from the steering wheel to operate the handbrake lever but finds it is 'not there'. This could increase the stress on the driver and possibly cause an accident. Levers for the operation of direction indicators, windscreen wipers, horn and lights are often mounted on the steering column but their positions are not standardised. It can be most frustrating and highly dangerous when a person who regularly drives a car, whose direction indicator lever is operated by the right hand, changes car and finds the same action results in the operation of the windscreen wipers.

Anatomy is the science of structure of the body. It is necessary for designers of machines to take into account the physical character-istics of the people who are to operate them. No two human beings are identical and the designer must allow for as wide a range of build as is mechanically possible. An operator should be able to reach his controls and switches easily without undue extra effort. If he operates from a sitting position his leg room, seat width and height of operation and viewing must be satisfactory.

It is important that sex and race are considered by designers of machines and equipment. Women are usually shorter than men and have less reach, and these factors should be recognised when equip-ment is being designed for use by either sex. The same aspect will have to be considered for machines which are to be operated by orientals for they are generally much shorter than westerners.

GUARDS AND COVERS

The safeguarding of machinery is a task which requires unremitting and careful attention at all times in the interest of industrial accident prevention.

Wherever possible, dangerous parts should be eliminated or effectively enclosed in the initial design of the machinery. If they cannot be eliminated, then suitable safeguards should be incorporated as part of the design and if this is not possible provision should be made for safeguards to be easily incorporated at a later stage.

Provision should be made to facilitate the fitting of alternative types of safeguards on machinery where it is known that this will be necessary because the work to be done on it will vary. Lubrication and routine maintenance facilities should be incorporated outside the danger area wherever practicable.

The use of automatic equipment to move components, materials and substances into and out of machine tools and process machinery should be favourably considered, as this not only reduces the hazards to persons at various operating points but also diminishes the risk of injury when materials would otherwise need to be moved manually.

While automatic feeding and removal devices have much to offer in preventing accidents to machine operators, they can create danger when any faults are being rectified by maintenance staff. Care should be taken to ensure that the use of these devices does not introduce further trapping hazards between the devices and parts of the machine or materials being processed.

In selecting an appropriate safeguard for a particular type of machinery or danger area it should be borne in mind that a fixed guard provides the highest standard of protection and should be used as far as practicable where access to the danger area is not required during normal operation of the machinery.

Fig. 4.3 shows a fixed guard on a press tool which also incorporates a hook-type feeding arrangement. Many press operations unavoidably require manual feeding but even in such cases it is often unnecessary for hands actually to enter the tool area. Special hand tools such as tongs and magnetic or pneumatic devices for manipulating the work-pieces in and out of the dies should be used wherever practicable. It is often considered that hand feeding by tongs or similar devices is primitive and leads to restrictions on production, but in fact operators can become expert in the use of such tools and be capable of maintaining an economic output. The fact that he does not have to place his hands in the danger zone has a psychological effect not always readily assessable and the operator's output can be more consistent.

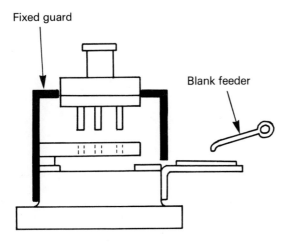

Fig. 4.3 A fixed safety guard and simple manual feeding arrangement. *(By courtesy of HMSO.)*

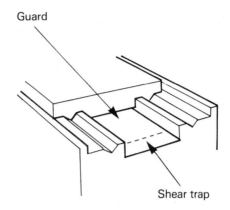

Fig. 4.4   A fixed safety guard for the bed of a planing machine.

Fig. 4.4 shows a guard for the bed of a planing machine. On some planing machines a trap is created as the machine table approaches the end of the bed, represented in the illustration by a broken line. The trap can be eliminated by means of a sheet metal guard, as shown, which extends beyond the inward travel of the table.

Fig. 4.5 shows a fixed guard constructed of wire mesh and angle section preventing access to transmission machinery.

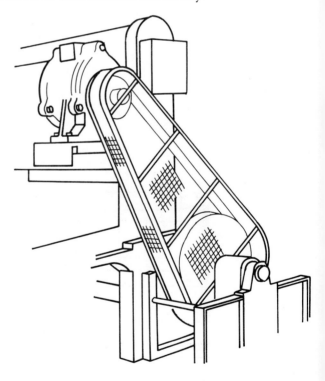

Fig. 4.5   A fixed safety guard to cover transmission machinery.

OTHER SAFETY DEVICES ———————

If all the dangerous parts of machinery and equipment were required to be made safe by fixed guarding it is certain that the commercial use of much useful plant would be prohibited. Fortunately there are available a large number of devices which are equally satisfactory in the quest for safe working. Some of these devices are now described.

Interlock Arrangements

An interlocking guard is one which has a moveable part which is connected to the machinery controls so that the part of the machinery causing danger cannot be set in motion until the guard is closed. Also access to the danger area is denied while danger exists.

The interlocking system may be either mechanical, electrical, hydraulic, pneumatic or any combination of these. The type of operation of the interlock should be considered in relation to the process to which it is applied. Figs. 4.6 and 4.7 show different types of interlock arrangements. Fig. 4.6 shows the principle of mechanical interlocking where a horizontally sliding guard is fitted to an upstroking hydraulic press. While the guard is open at (a) the lower edge of the guard retains the lever of the control valve in the safe position. When the guard has been closed (b) the lever can be moved to set the press in motion and the lever then locks the guard closed.

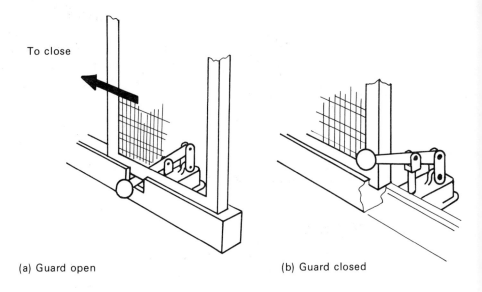

To close

(a) Guard open   (b) Guard closed

Fig. 4.6   Mechanical interlocking.

Fig. 4.7(a) and (b) show hydraulic valves which may be interlocked with one another by the specially shaped plates P and Q fitted respectively to the exhaust and pressure valves.

In Fig. 4.7(a) the exhaust valve P is closed and the pressure valve Q is shown open. Fig. 4.7(b) shows the exhaust valve P open and the pressure valve Q locked closed. When the opening and closing of a sliding guard is incorporated in this arrangement it forms an efficient method of interlocking, Fig. 4.7(c). At (c) a sliding guard R fitted on runners carries an extension S. The exhaust valve P is fitted with an integral boss in which is a slot of sufficient width to accommodate guard extension S. In the position shown the exhaust valve P is open and the slot in the boss is ready to accommodate the guard extension S when it is opened by sliding it to the left. On completion of this action the pressure valve Q cannot be opened and the exhaust valve P cannot be closed. A stop (not shown) should be provided to ensure that the slot in the boss is brought into the correct position each time the valve is operated.

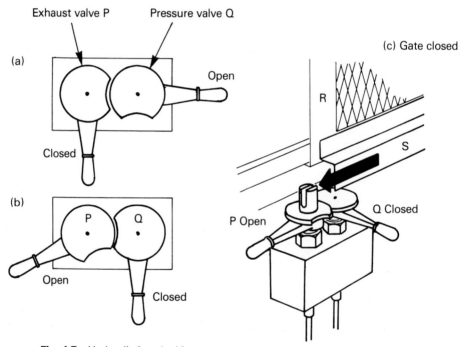

Fig. 4.7 Hydraulic interlocking.

Safety Valves

Boilers are designed to sustain a certain amount of internal pressure. If the design pressure is exceeded for any length of time, there is a risk that the pressure may increase until there is an evident chance of explosion.

To prevent this pressure build-up a safety valve is fitted to the boiler. A typical safety valve is shown in Fig. 4.8. The compression spring forces the valve on to its seat to form a seal and the boiler will operate at its normal working pressure. If the boiler pressure is increased towards the danger level, it will lift the valve from its seat compressing the spring and steam will exhaust reducing the pressure inside the boiler to an acceptable level.

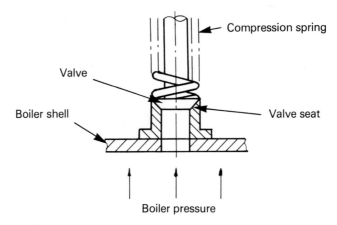

Fig. 4.8   A safety valve.

Fuses

A fuse is used to protect an electric circuit from an overload of current. The amount of current that can pass through a conductor is limited. Beyond this limit, heat develops and damage to the insulation occurs, causing a breakdown. All domestic and industrial circuits are fitted with fuses to prevent such overloading, which can cause a fire.

A fuse is a deliberate weak spot designed to fail as soon as the circuit becomes overloaded. It is vital always to fit a fuse of the correct rating for a particular appliance or circuit. Never replace a fuse with another of higher rating or with any substitute piece of metal. A domestic fuse is illustrated in Fig. 4.9.

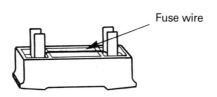

Fig. 4.9   A domestic fuse.

Governors

The purpose of a governor is to control the speed of an engine. In the Watt governor, illustrated in Fig. 4.10, rotating steel balls are suspended by links which are pivoted on the axis of rotation. The speed of rotation of the central spindle is maintained directly from the engine. When the balls are rotating the centrifugal force acting on them causes them to move outward and upward as the speed increases lifting the sleeve higher up the spindle. By means of a suitable linkage, the movement of the sleeve closes the accelerator valve and so reduces the engine speed until it falls to its normal value.

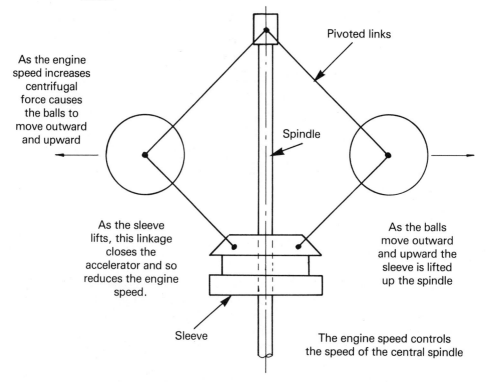

As the engine speed increases centrifugal force causes the balls to move outward and upward

Pivoted links

Spindle

As the sleeve lifts, this linkage closes the accelerator and so reduces the engine speed.

As the balls move outward and upward the sleeve is lifted up the spindle

Sleeve

The engine speed controls the speed of the central spindle

Fig. 4.10 The Watt governor.

Warning Lights

As the name implies these lights are used as a warning to tell the operator that a certain malfunction is occurring or action needs to be taken.

Figs. 3.3 and 3.6 show a motor car facia panel where warning lights are used to indicate that various elements are being used, e.g. handbrake warning light. This light warns the driver that the handbrake is still on and failure to release it correctly may lead to an accident.

In some cases where the warning may be most urgent the light is made to flash intermittently in order to draw the immediate attention of the operator. Examples of this can also be seen in Figs. 3.3–3.5, e.g. warning lights for low fuel, direction indicator and hazard warning.

Buzzers

Vision is a special sense which is most easily overloaded. Sometimes it is necessary to reinforce the visual display, say of a variation which goes past a limiting condition, by means of an alarm to attract the attention of the operator through another of his senses. Most frequently an audible signal is used, such as a buzzer.

FAIL-SAFE SYSTEMS AND DEVICES

Where danger may arise a safety device should be incorporated in the design to restrain any moving machinery part which may overrun its normal stopping position due to gravity, the fall back of a crankshaft or the failure of a counterweight system.

All such devices should fail to safety. They must stop and reverse the movement of the dangerous parts before injury can occur or the safeguard must remain in position to prevent access to the danger area.

Brakes are used in order to stop machinery quickly. Braking systems should be so designed as to bring dangerous moving parts to rest with a consistent performance as quickly as possible. Wherever practicable, all braking systems should fail to safety, i.e. if the power supply is interrupted the brake is applied.

Counterweights are used to reduce the effort required to open a rise-and-fall guard. An anti-fall stop device should be incorporated in such a system to arrest the movement of a falling guard in the event of breakage of the supporting cable thus allowing a failure to safety.

Fig. 4.11 shows a progressive mechanical restraint device suitable for a pressure die-casting machine or plastics injection moulding machine. Bar A is attached to the moving die platen and the restraint B, which pivots about C and is controlled by air cylinder D, is mounted on the fixed part of the machine. When the guard is closed restraint B is disengaged allowing the moving platen to close. While the platen is opening the projection on restraint B is clear of the projection on bar A. If, due to a fault in the electrical or hydraulic circuits, the platen starts to close again before it has reached its fully open position, the restraint will arrest the closing movement before trapping can occur. Because of the design of the electrical, hydraulic

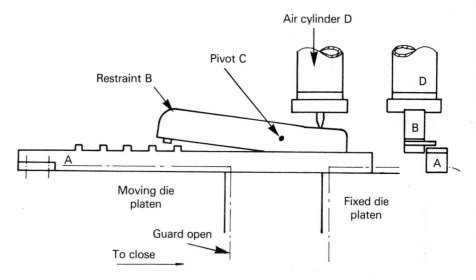

Fig. 4.11 A progressive mechanical restraint device.

and pneumatic circuits, a failure in the electrical or hydraulic circuits will result in the piston rod of the air cylinder retracting and, because of gravity, restraint B will fall allowing its projection to locate in one of the projections on bar A and prevent the closing of the platen. In the event of failure of the air supply to cylinder D the piston rod will retract and restraint B will again engage by gravity.

A safety hook, to protect against gravity fall of an air-operated guard which has no balance weight, is shown in Fig. 4.12. During operation

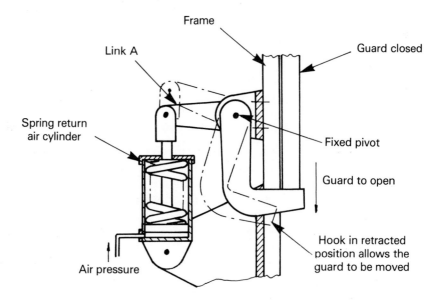

Fig. 4.12 A safety hook on an air-operated guard.

the hook is actuated by a single acting spring return air cylinder, air pressure being used to disengage the hook. In the event of air pressure failure the spring forces the piston to the bottom of the air cylinder actuating the hook through link A and the guard cannot be moved and access is prevented to the danger area.

QUESTIONS

1. Under the terms of the Health and Safety at Work, etc. Act, 1974, what are the duties of a designer?

2. What are 'statutory duties'?

3. What considerations must be made in order that a designer can evaluate the balance between safety and design requirements of a product?

4. Give an example of how each of the following aspects affect man, machine and his working environment:
   - (i) physiological;
   - (ii) psychological;
   - (iii) anatomical.

5. Which type of guard gives the best protection?

6. What particular design care needs to be made in order to protect maintenance and tool-setting staff on machines with automatic feeding and removal devices?

7. State the use and, from your own workshop, give an example of three of the following safety devices:
   - (i) interlock arrangement;
   - (ii) safety valve;
   - (iii) fuse;
   - (iv) warning light;
   - (v) buzzer.

8. Sketch and describe the operation of a Watt governor.

9. Describe the concept of the fail-to-safety system.

10. Sketch and describe the operation of a fail-safe device in your own workshop.

5 Design of Simple Link and Rotary Mechanisms

After reading this chapter you should be able to:

★ describe the design of simple link and rotary mechanisms (G);

★ sketch and describe examples of simple link mechanisms for converting one type of motion to another, without an analysis of velocities or accelerations (S);

★ sketch and describe simple gear, pulley and screw mechanisms for converting one type of rotary motion to another, without an analysis of velocities or accelerations (S).

(G) = general TEC objective
(S) = specific TEC objective

INTRODUCTION

A mechanism is a device constructed from a number of moving parts which transmit motion.

When designing mechanisms for a specific function, the designer makes use of standard constructional elements such as links, gears, pulleys, screws and nuts.

SIMPLE LINK MECHANISMS

A simple link mechanism may be used to convert reciprocating straight-line motion into an equal and opposite motion. Fig. 5.1(a) shows a link mechanism where the movement of rod X, which is limited to straight-line motion due to the constraints of the guide, causes the link L to move about pivot P by the thrust of the slider pin S on the link slot. A slot is required in the link to allow for radial displacement during movement. The movement of the link causes the slider pin T to thrust against the slot in the other end of the link resulting in the movement of rod Y in the opposite direction to the movement of rod X.

Fig. 5.1(b) shows a similar link mechanism where a pulling force is applied to rod X. The resulting movement of rod Y is equal and opposite to the movement of rod X.

In both cases, if the movement of rod X is reversed after it has reached the end of its stroke, thus imparting reciprocating motion, equal and opposite reciprocating movement will be given to rod Y.

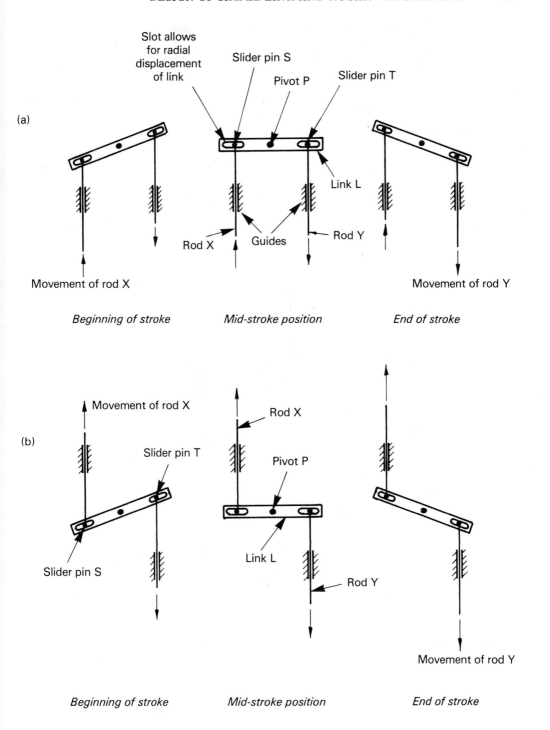

Fig. 5.1 Simple link mechanisms used to reverse the direction of motion.

Fig. 5.2 shows the bell crank lever system which may be used to transmit reciprocating straight-line motion into an equal motion displaced through 90°. As in the previous example the movement of

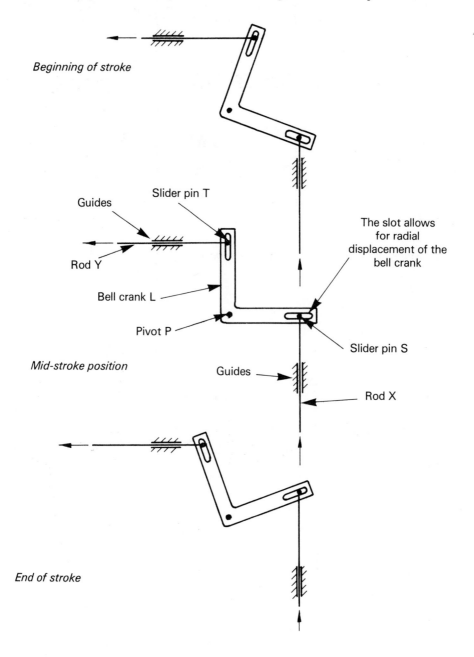

Fig. 5.2   The bell crank used to transmit motion through 90°.

rod X, again constrained by a guide, causes the slider pin S to move the bell crank L about pivot P effecting the movement of rod Y through slider pin T. Again if the movement of rod X is reversed after it has reached the end of its stroke, giving reciprocating motion, equal and opposite reciprocating movement will be imparted to rod Y.

Another way of converting reciprocating straight-line motion into motion displaced through 90° is by use of the wedge cam illustrated in Fig. 5.3. Because of their relative simplicity cam systems are widely used for reciprocating and intermittent motion. As the cam moves to the left, the roller follower, which freely rotates about its own centre, is forced to rise due to the taper form of the wedge and movement is imparted to the rod R. The total movement of the rod is determined by the design of the amount of rise of the cam. Fig. 5.3(a) shows the cam just after the beginning of its stroke and Fig. 5.3(b) shows it almost at the end of its stroke.

If the movement of the cam is reversed after it has reached the end of its stroke, giving reciprocating motion, the movement of the roller follower will be reversed with the help of the action of the compression spring and reciprocating movement will be imparted to the rod.

Fig. 5.4 shows a disc cam which is used to convert radial movement into linear motion. As the cam rotates about centre O the radial displacement, the rise of the cam, causes the roller follower and the rod R to rise. When the roller follower has reached the maximum radial displacement position of the cam, as shown in Fig. 5.4(b), it will remain stationary in this instance for the shape of the cam is truly radial about centre O for a further 120° of rotary movement. This is called the 'dwell position'. Thereafter the reduction in radial displacement, the fall of the cam, causes the roller follower and thus the rod to move back to the initial position shown in Fig. 5.4(a), helped by the action of the compression spring. In one complete revolution of the cam the rod R will rise, dwell and fall giving intermittent reciprocating motion.

Special or irregular motion in a follower can be obtained depending upon the design of the cam profile.

THE CRANK AND CONNECTING ROD

In the internal combustion engine, the ignition of the vaporised fuel in the combustion chamber causes an explosion which imposes a thrust on the piston. The straight-line movement of the piston is transmitted through the connecting rod linkage, and causes the crank to rotate about the crankshaft. It should be noted that in the internal combustion engine the crank and crankshaft are usually one integral unit.

(a)

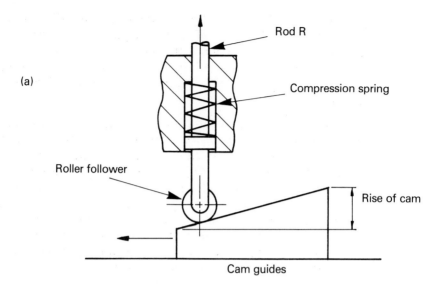

Rod R

Compression spring

Roller follower

Rise of cam

Cam guides

(b)

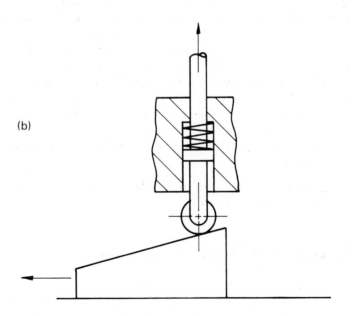

Fig. 5.3 Motion transmitted through 90° using a wedge cam.

(a)

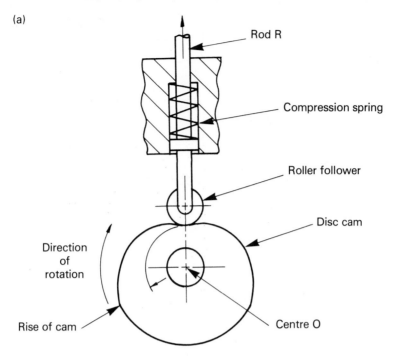

Rod R

Compression spring

Roller follower

Disc cam

Direction
of
rotation

Rise of cam

Centre O

(b)

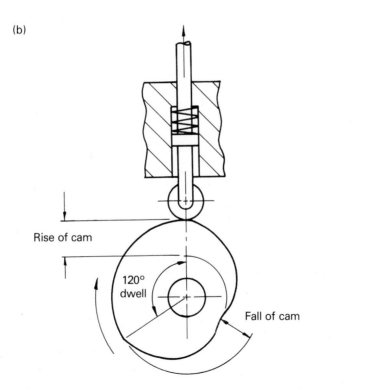

Rise of cam

120°
dwell

Fall of cam

Fig. 5.4 A disc cam used to translate a radial displacement (rise) into a linear
displacement.

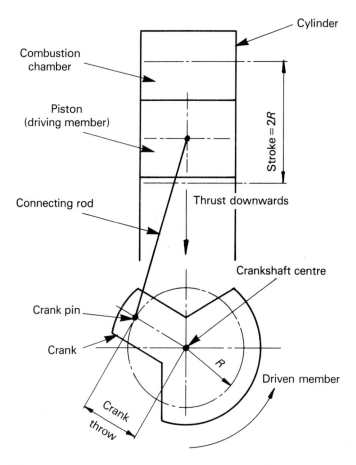

Fig. 5.5 Rotary motion obtained from straight-line motion in the internal combustion engine.

A line diagram of this mechanism is shown in Fig. 5.5. This movement is shown in skeletal form only. Since so much preliminary work in machine design involves just kinematic consideration, simple structural diagrams, such as this, are all that is necessary. A model of the internal combustion engine is shown in Fig. 5.6.

In many machine tool applications straight-line or reciprocating movement is required, therefore the converse of the above principle is used, as shown in Fig. 5.7. The rotary movement provided by the electric motor causes the crank to rotate about centre O and the crank pin follows the circumference of the pitch circle causing the slider block, which is attached to the crank pin through a connecting rod and pin joint, to move in a straight line due to the constraint of the guides. While the rotating crank is in the upper arc of its movement the connecting rod and slider block move in a forward direction. When the crank pin has reached position B the straight-line

Fig. 5.6   A model of the internal combustion engine showing the piston, connecting rod and crankshaft.

motion is reversed and the connecting rod and slider block move in a backwards direction. When the crank pin reaches position A the straight-line motion is reversed once again thus giving reciprocating movement. It should be noted that the longer the connecting rod becomes the nearer the motion is to simple harmonic motion and the smaller the force on the guides. In the Scotch yoke mechanism (Fig. 5.8), the movement *is* simple harmonic.

In this mechanism the rotary movement of an electric motor causes the crank pin, which is affixed to the crank, to rotate about centre O. The adaptor, which is connected to the crank pin and rotates with it, causes the crosshead and the slider block, which are firmly attached to each other, to move horizontally. The slider block thus has the same motion as the projection of the crank pin on the horizontal diameter.

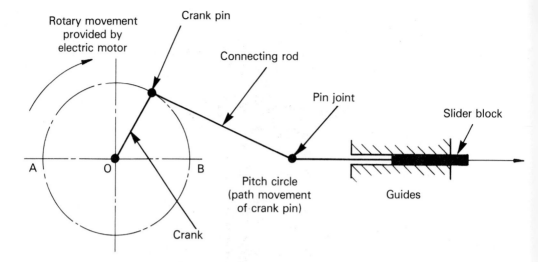

Fig. 5.7 Straight-line motion obtained from rotary motion.

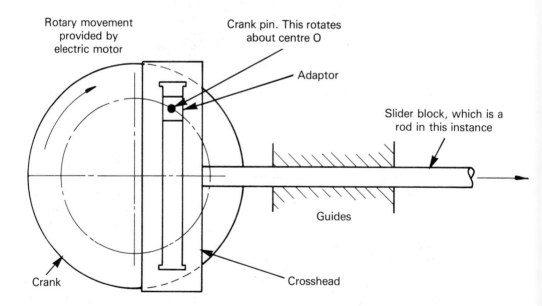

Fig. 5.8 The 'Scotch yoke' mechanism, eliminating the irregularities in motion associated with the crank and connecting rod.

This mechanism is very compact compared with the crank and connecting rod (Fig. 5.7).

Let us now consider some applications of the crank and connecting rod mechanism.

The Slotted-link Mechanism

The slotted-link mechanism (Fig. 5.9) is used to obtain reciprocation of the ram on a shaping machine and also achieves a quick-return motion.

The crank A, sometimes called the bull gear, revolves about centre O. The crank pin X moves, with adaptor B, in the slot of the link, which moves about a pivot, centre C. The extreme positions of the other end E of the link occur when the centre line of the link is tangential

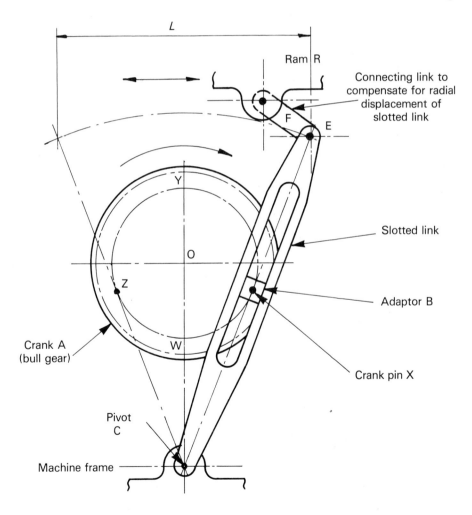

Fig. 5.9 The slotted-link mechanism.

to the crank circle, i.e. when the crank pin is at X and Z. Thus the end E moves over an arc of a circle corresponding to the horizontal distance *L*. E is connected to the ram R by link F which compensates for the arc movement of E allowing the ram to be constrained to move over a horizontal surface and the cutting tool, which is located on the head at the front of the ram, also moves a horizontal distance *L*, to and fro, as A rotates.

Motion of the tool from left to right occurs as the crank pin moves over arc ZYX; return motion as it moves over arc XWZ. As A rotates uniformly and XWZ is less than ZYX, it follows that the time for the return stroke is less than that for the forward stroke. This principle is shown in skeletal form in Fig. 5.10(a).

The length of stroke of the ram may be varied by altering the radial position of the crank pin X (Fig. 5.10(b)). Altering the length of stroke also alters the cutting time/return time ratio. This can be seen by comparing the angles moved through by the crank pin on the cutting stroke and return stroke (Figs. 5.10(a) and (b)).

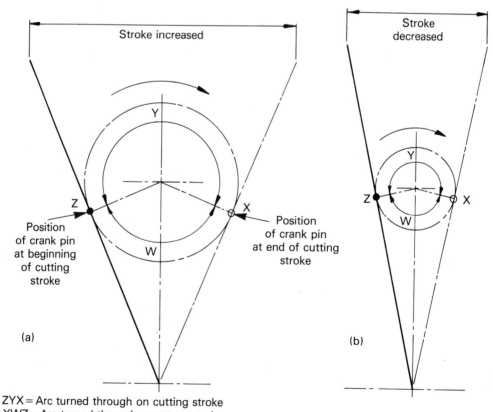

ZYX = Arc turned through on cutting stroke
XWZ = Arc turned through on return stroke

Fig. 5.10 The motion of a slotted-link mechanism.

The Shaping Machine Feed Mechanism

When generating flat surfaces on a shaping machine it is necessary to feed the work beneath the reciprocating cutting tool. This is achieved by indexing a feed shaft.

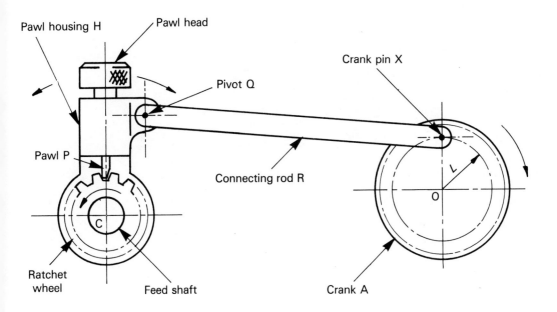

Fig. 5.11   The shaping machine feed mechanism.

The mechanism for this feed system is shown in Fig. 5.11. The crank A revolves about centre O. Crank pin X rotates with the crank and the connecting rod R causes pivot Q, which is constrained to move in an arc about feed shaft centre C, to move the pawl housing H to and fro. As the connecting rod is moved to the right the spring loaded pawl P rides over the teeth on the ratchet wheel, due to the chamfer on its advancing face, and the feed shaft, which is firmly attached to the ratchet wheel, remains stationary. As the connecting rod moves to the left the pawl pushes on to a tooth on the ratchet wheel and causes it to rotate slightly, indexing the feed shaft by a similar angular amount, thus imparting a feed to the work table.

As radius L is reduced the corresponding movement of the pawl is decreased to give a consequent reduction in feed. The direction of feed may be reversed by lifting and turning the spring-loaded pawl through 180°.

Fig. 5.12 shows the feed mechanism on the shaping machine.

Fig. 5.12   The feed mechanism arrangement on the shaping machine.   *(By courtesy of the Victoria Machine Tool Company.)*

The Power Hacksaw

The power hacksaw (Fig. 5.13) is used for cutting off bar stock of various sections. Unfortunately this machine has a discontinuous cutting action. It cuts in one direction only, and in some cases is being replaced by the bandsaw type machine.

The crank disc is rotated by an electric motor and belt drive causing the crank pin to rotate and allow the frame to reciprocate, in the overarm slides, through the connecting rod. The overarm is pivoted to accommodate loading of the machine.

The Whitworth Quick-return Mechanism

A modified crank and connecting rod mechanism, known as the Whitworth quick-return mechanism, is used to impart reciprocating motion to the vertical ram of the slotting machine. A representation of this system is shown, in skeletal form, with the ram in three operating positions in Fig. 5.14.

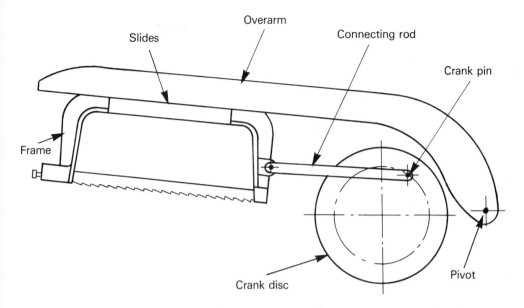

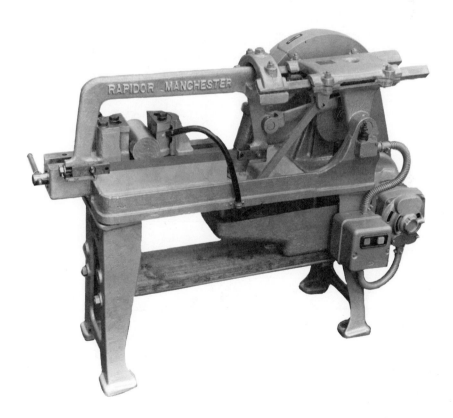

Fig. 5.13 The power hacksaw: (a) schematic diagram; (b) photograph *(By courtesy of Alexander Machinery Ltd.)*

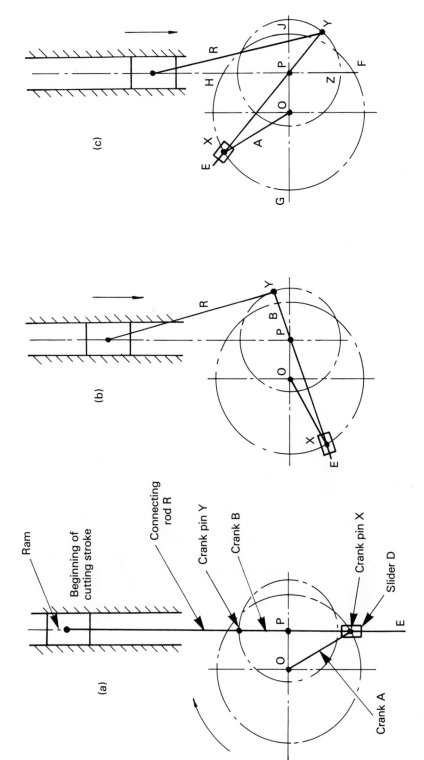

Fig. 5.14   The Whitworth quick-return mechanism.

Fig. 5.14(a) shows the ram at the beginning of the cutting stroke. The crank A revolves about centre O. The crank pin X rotates with slider D and causes crank B, and its extension to point E, to revolve about centre P. Crank pin Y rotates and, through the connecting rod R, causes the ram to descend.

Fig. 5.14(b) shows the ram during its cutting stroke, the extension of crank B, at E, has moved in the slider, lengthening the distance from P to X.

Fig. 5.14(c) shows the ram approaching the end of the cutting stroke, which will occur when crank pin Y arrives at position Z and crank pin X will be at position H. The cutting stroke will have taken place while crank A is rotating through the arc FGH and the return stroke through arc HJF. As the crank will be rotating at a constant rate, the return stroke will be quicker.

The length of stroke of the ram is varied by altering the radial position of crank pin Y.

The Toggle Link Mechanism

The toggle link mechanism is widely used in clamping devices. It may be operated by hand or by mechanical, hydraulic or pneumatic means. Fig. 5.15 shows a pneumatically operated clamping device. During the forward stroke of the air cylinder piston the links are moved towards the stop causing the moving jaw to clamp the work.

An advantage of this .mechanism is that as the links approach a straight line the clamping force is greatly increased.

The links may be locked by moving slightly over centre against a stop. It should be noted that this movement over centre is greatly exaggerated in Fig. 5.15(a) and in practice this movement is very small indeed.

Fig. 5.16 shows a hand-operated toggle clamp. The handle H pivots about centre O. As it is moved forward (to the right) link L causes the clamping arm A, which pivots about centre C, to move in an arc. As the link pins, O, X and Y approach a straight line the clamping force is greatly increased. The design allows this to occur as the clamping arm reaches the horizontal position.

GEAR, PULLEY AND SCREW MECHANISMS ————

Gear Mechanisms

Gears are used to transmit rotary motion from one shaft to another and are used when a positive drive is required. Their speed ratio is

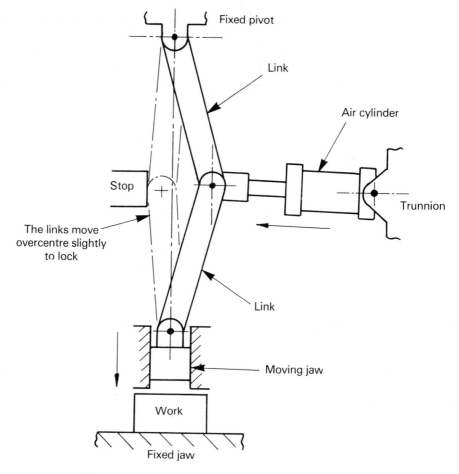

Fig. 5.15 A pneumatically operated toggle link mechanism.

Fig. 5.16 A hand-operated toggle clamp. Notice the position of the three link pins. As the handle is pushed forward the clamp closes and the pins take up a common centre line, greatly increasing the clamping force.

determined by the numbers of teeth on the gears so it is possible to maintain an exact relationship between two mating gears. Fig. 5.17 shows two spur gears in mesh. Their direction of rotation will be opposite to one another.

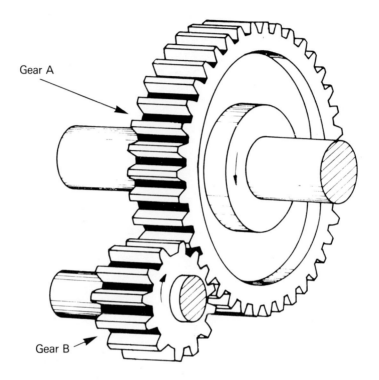

Gear A

Gear B

Fig. 5.17 Spur gears.

The number of teeth on gear A is 39 and the number of teeth on gear B is 13. As gear A makes 1 revolution gear B will make $\frac{39}{13} = 3$ revolutions.

If the two gears must rotate in the same direction, an idler gear, of any size, is fitted between them. This idler gear does not affect the speed ratio between the driver gear A and the driven gear B (Figs. 5.18 and 5.19).

Bevel gears are used to connect shafts which lie at an angle to each other (often at right angles to each other) (Fig. 5.20). The speed ratio is determined in a similar way to that of spur gears.

Spur gears are used in lathe gear trains and gear boxes, and bevel gears are often used to raise the table of a milling machine.

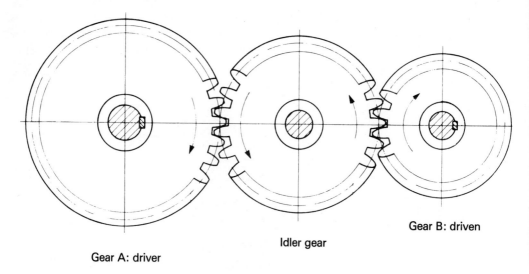

Gear B: driven

Idler gear

Gear A: driver

Fig. 5.18   The use of an idler gear to let gears A and B rotate in the same direction.

Worm and Worm-wheel

Worm gearing is another method of driving two shafts which lie at right angles to each other (Fig. 5.21). One revolution of the worm will cause the worm-wheel to rotate through one tooth. This type of mechanism is ideal where a large speed ratio is required, such as in a dividing head.

Rack and Pinion

The rack and pinion (Fig. 5.22), is a system of gearing which converts rotary motion into straight-line motion. Each tooth of the pinion engages with one tooth on the rack so that one revolution of the pinion gives a straight-line movement equal to πd (where d is the pitch circle diameter of the pinion). This principle is used to traverse a lathe saddle along the bed.

Pulley Mechanisms

Rotary motion may be transmitted from one shaft to another by means of belt and pulley systems (Fig. 5.23). The speed ratio is determined by the pulley diameters. If pulley A is 300 mm diameter and pulley B is 150 mm diameter, in one revolution of pulley A, pulley B would rotate $\frac{300}{150} = 2$ revolutions.

Gear A (15 teeth is rotated from the main spindle through the gear box in the headstock). It is the driver.
Gear B (100 teeth) is the idler gear. The number of teeth on this gear does not affect the ratio A:C (A and C rotate in the same direction).
Gear C (80 teeth) is the driven gear controlling the rotation of the leadscrew and feedshaft.
Gear P is rotated from the gearbox and drives gear Q.
Gear Q is the main spindle gear and is driven by gear P.
As gear A makes 8 revolutions, gear C revolves $8 \times \frac{15}{80}$
$= 1.5$ revolutions.

Fig. 5.19   A simple gear train from the main spindle to the leadscrew and feedshaft of a centre lathe.

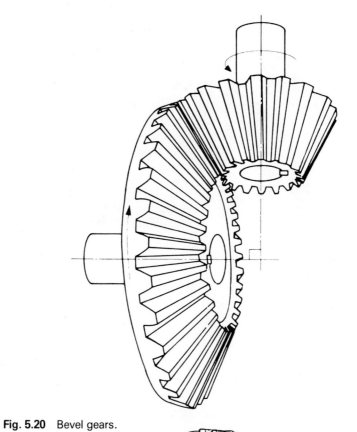

Fig. 5.20   Bevel gears.

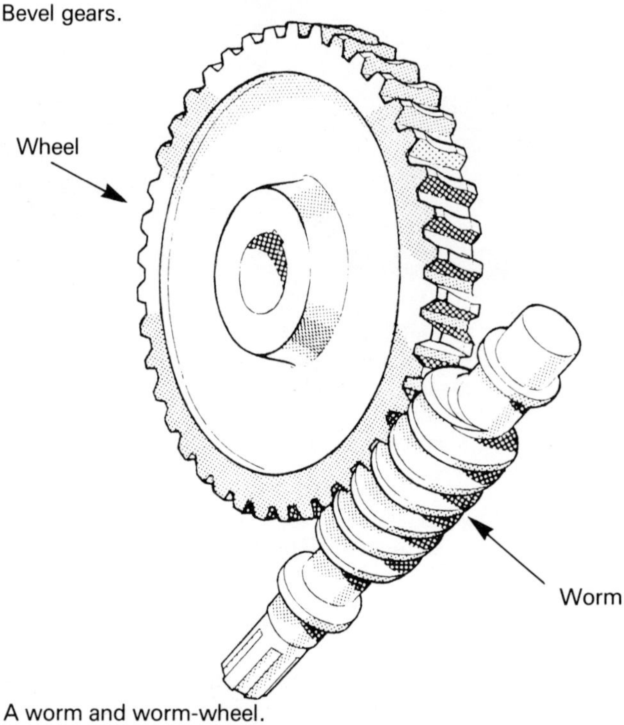

Wheel

Worm

Fig. 5.21   A worm and worm-wheel.

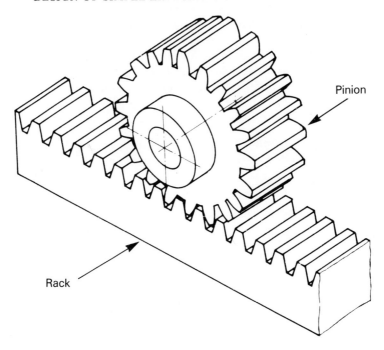

Fig. 5.22   A rack and pinion.

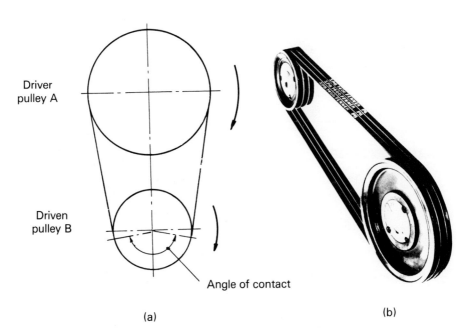

Driver
pulley A

Driven
pulley B

Angle of contact

(a)

(b)

Fig. 5.23   A belt drive: (a) schematic;   (b) photograph *(By courtesy of J H Fenner and Co. Ltd.)*

In a belt drive power is transmitted by friction between belt and pulley; hence a positive drive is not attained and slip may occur. This type of mechanism has the advantage over a gear mechanism when the distance between driver and driven shaft is great. It has a disadvantage, however, where the shaft centre distance is small, the angle of contact between belt and pulley is reduced hence increasing the chance of belt slip.

If two pulleys must rotate in opposite directions the belt may be crossed (Fig. 5.24).

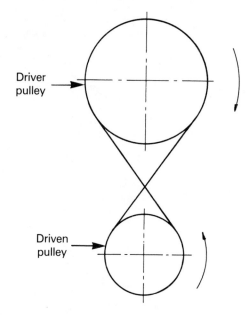

Driver pulley

Driven pulley

Fig. 5.24   Crossed belt drive.

Chain Drives

Rotary motion may also be transmitted from one shaft to another by means of roller chain drives (Fig. 5.25). The speed ratio is determined by the ratio of the mean diameters of the chain wheels and a positive drive is achieved.

Example   If a driving wheel of a chain drive, which has a mean diameter of 75 mm, is rotating at 50 rev/min, determine the speed of the driven chain wheel, which has a mean diameter of 125 mm.

Solution

$$\text{Driven speed} \ = \ 50 \, \text{rev/min} \times \frac{75}{125} \ = \ 30 \, \text{rev/min}$$

Fig. 5.25   A chain drive.   *(By courtesy of J H Fenner and Co. Ltd.)*

Nut-and-screw Mechanisms

A nut-and-screw mechanism is utilised in a screw jack (see Fig. 5.26(a). The screw jack consists of a heavy base, with a central vertical hole, with a square thread. Into this hole fits a threaded spindle with a head (to support a load) and a handle. One turn of the handle raises the load by an amount equal to the lead of the screw.

Fig. 5.26(b) shows a screw jack which is used for experimental purposes in a laboratory. As it is only used for light applications, a handle is not fitted as the head can easily be turned by hand.

A nut-and-screw mechanism is also incorporated in the movement of slideway systems found on most machine tools. A cross-slide system (Fig. 5.27) utilises this type of mechanism. The machine is provided with a square-threaded hardened lead screw having a handwheel at one end for the operator and a shank and collar such that when the thread is rotated, no axial movement is possible. On the thread is a phosphor-bronze nut recessed into the moving member (the cross-slide, in this figure). Turning the handwheel moves the nut, and therefore the slide, backwards or forwards, according to the direction of rotation of the handwheel. By calibrating a dial on the handwheel, one can measure the movement of the slide.

Fig. 5.28 shows how this system can be used in the cross-slide and compound slide of a centre lathe.

A combination of gear and lever mechanisms can be seen in Fig. 5.29, which details a motor car rack-and-pinion steering arrangement.

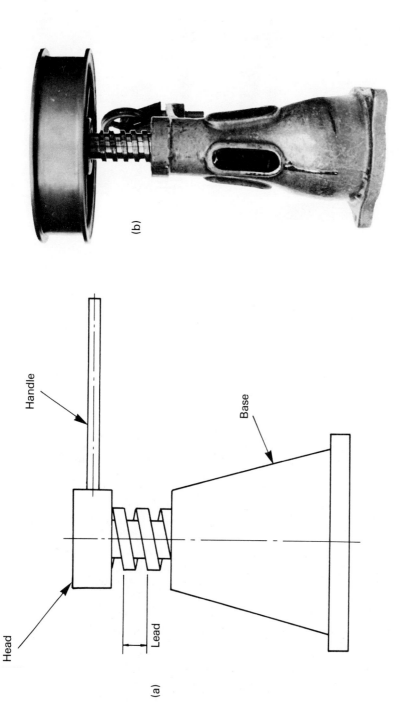

Handle

Head

Lead

Base

(a)

(b)

Fig. 5.26 A screw jack: (a) schematic;   (b) photograph.

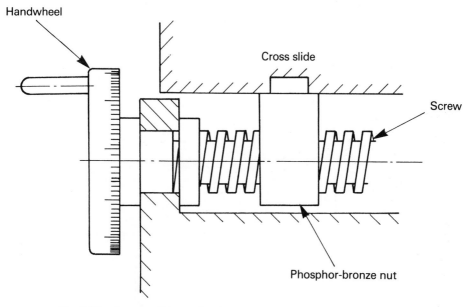

Fig. 5.27 A cross-slide mechanism.

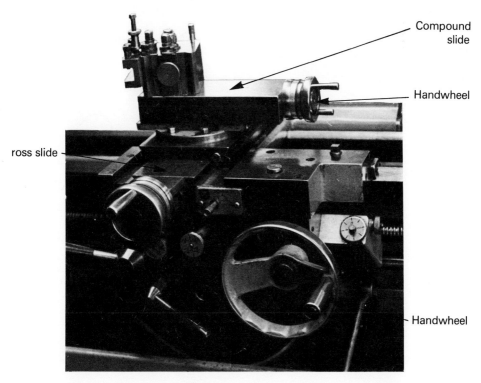

Fig. 5.28 The screw-and-nut feed arrangement used in the cross slide and compound slide of a centre lathe.

The rotational movements of the steering wheel are converted into linear movement by a rack-and-pinion, as illustrated at Fig. 5.22. This movement is used to turn the wheels to left or right by means of the pin joints shown.

Fig. 5.29 shows the steering gear in the central position (a) and in the left-hand full lock position (b). It can be seen that the rack has reached its limiting position.

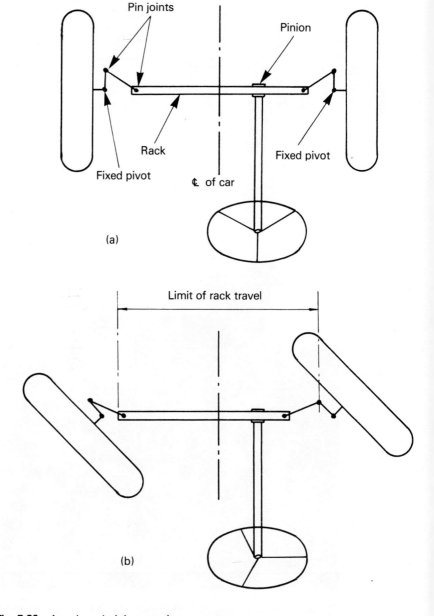

Fig. 5.29   A rack-and-pinion steering arrangement.

QUESTIONS

1. State two reasons why it is desirable to have the connecting rod as long as possible in the crank and connecting rod mechanism.

2. Sketch an alternative method to the one of using a connecting link in Fig. 5.9 to allow for radial displacement of the slotted link in the shaping machine quick-return mechanism.

3. Sketch a simple link mechanism for transmitting straight line motion through 90°.

4. Using a sketch describe how the length of stroke is governed in the crank-and-connecting-rod mechanism.

5. State one reason why the toggle link mechanism is used in clamping devices.

6. Sketch a mechanism suitable for converting the clockwise rotary motion of a driver at 12 rev/min to anticlockwise rotary motion at 4 rev/min. Mark on your sketch the driver and driven elements.

7. Sketch two separate mechanisms for connecting shafts at right angles. Give one example where each system is used in engineering.

8. Fig. 5.28 shows a nut-and-screw mechanism, used for a lathe, where the nut travels with the cross-slide. Sketch a similar system where the nut is in a fixed position and the screw travels with the cross-slide.

9. Sketch two separate mechanisms for converting rotary motion into linear motion, and give an example where each system is used in engineering.

10. State two disadvantages of the belt drive system for transmitting power.

Preparation and Evaluation of Designs

A CASE STUDY IN DESIGN

When a designer is presented with a design requirement he must determine exactly just what condition needs to be satisfied. A designer is a person who solves problems. He must decide the essential qualities of the design and examine the various factors which will affect it. Quite often there is no *one* solution which immediately presents itself as the answer. A designer will develop ideas with a number of possible solutions — he will synthesise — and select the best method by comparison and evaluation.

The cost of production will be a major factor in his considerations, for ultimately his design will need to be manufactured. If the product is expensive to produce, its manufacture may not be viable. Consider the design of a new motor car, which is basically a body shell, with seats for comfort and protection, a driving unit and four wheels. When a car designer is developing a new model, he must be vitally aware of the costs of production. He cannot include the refinements and quality of a custom-built saloon at the same cost as a production-line model.

The first stage of a design plan is to find out the true nature of the problem, and a major factor influencing this is the way the problem is defined. A considerable amount of valuable time may be lost, in evolving the design, by not defining the problem accurately and hence allowing the project to start off from the wrong base.

Once the problem has been satisfactorily defined, all the factors which influence the design can then be examined. These factors

192

are the requirements of the design — just what is the design expected to achieve? From these requirements a *specification* can be written.

A specification is a method of conveying the requirements of the design to all the people involved in it. It will include the precise details of the demands of the design and will state the limits, if any, in physical size and mass of the product and the forces applied to it or resulting from its action. The type of material and protective finish will be included if they need to be of a specific nature. The student is advised to look at a helpful booklet, published by the British Standards Institution, called *Guide to the Preparation of Specifications, PD 6112*, which details the items which should be included in specifications.

The next stage of the design plan is to compare a number of possible solutions, evaluate each one, and select the most suitable. Solutions may be evolved by freehand sketching. This method of design can be quite satisfactory for selection purposes and will be less time-consuming as a result. The final solution may be a compromise between requirements which are in contention with each other and for this reason it is often better for an individual to select the most suitable solution after a team of designers has presented its observations.

When the final design decision has been made, the design sketch has to be translated into working drawings to facilitate the manufacture of the component. This work is usually done by detail designers and draughtsmen and may be of considerable content. There are often many calculations and some analysis still to be undertaken before the drawings are completed.

The system of production of working drawings does vary between different drawing offices. The most common method is to start to make an assembly drawing until something like the final design can be appreciated, produce the detail drawings from this, and then make a complete assembly drawing from the detail drawings.

As a case study, let us now consider the requirements for a particular component, make a number of designs and, by the procedure previously outlined, arrive at a final design by comparison and evaluation.

There is an urgent requirement for sliding door suspension units suitable for garages in a new housing development. 1200 houses are being built by a large construction company, RCT Ltd, who had previously accepted a quotation for the supplying and fitting of the suspension units and track. Unfortunately, after delivery of the track, the supplier, who was in financial difficulties, went into liquidation and ceased trading. The builders now have the immediate problem of finding a company who can design and manufacture suspension units to fit an existing enclosed track. They have invited tenders for the supply of an initial batch of 9000 units with the possibilities of repeat orders for their other sites.

Messrs Brokham Bros, a local engineering company, who are keen to expand their interests and become suppliers to the building industry, have arranged for their Sales Manager and Chief Designer to have preliminary discussions with RCT Ltd. They have ascertained that the development will consist of 900 detached and semidetached dwellings, each with a single garage (5 units per garage), and 300 superior-type dwellings, each with a double garage (11 units per garage).

The designer, with his small design team, have considered all the factors which will affect the design and collectively made a list of the requirements of the suspension units. They have written the following specification.

Specification for Sliding Door Suspension Units

1. Foreword

    This specification is made to suit the requirements of RCT Ltd. for sliding door suspension units suitable for garage doors in domestic dwellings.

2. Scope

    This specification applies to the sliding door suspension units only which must be compatible with the existing low-carbon steel rolled section track (see Fig. 6.1).

3. Definitions

    (a) Abbreviations:

    Unit = Sliding door suspension unit

    (b) Measuring system:

    Metric units are to be used throughout; masses to be expressed in kilogrammes (kg) and forces to be expressed in newtons (N). The Detail and Assembly drawings to be in accordance with BS.308: 1972

4. Conditions

    (a) The unit will operate in a temperature range of $-5\,^{\circ}C$ to $30\,^{\circ}C$.

    (b) The unit will operate in a dust and grit-laden atmosphere.

    (c) Grease will be liberally applied between the mating surfaces of unit and track.

    (d) The only routine maintenance that will be done will be to annually wipe down the unit and track, to remove dust and grit, and re-grease.

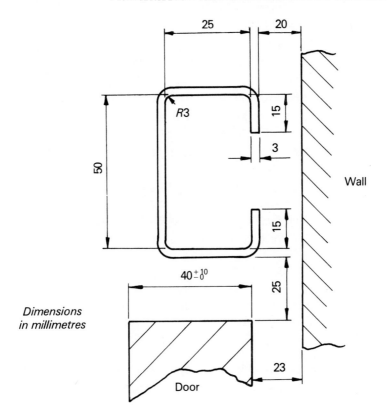

*Dimensions
in millimetres*

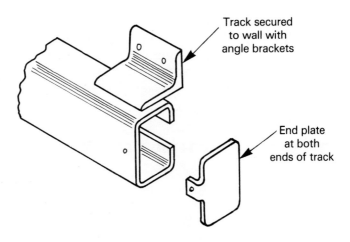

Fig. 6.1   Details of a sliding door track.

5. Characteristics

(a) All parts of the unit which will come into contact with hands during assembly or maintenance shall have no sharp edges.

(b) The units are to be used on wooden doors.

(c) Each wooden door will have a height of 2.10 m, a width of 800 – 900 mm and a thickness of 40 – 50 mm

(d) The mass of each door will be a minimum of 40 kg and a maximum of 50 kg.

(e) During operation the doors will slide along the track and down the side of the garage, turning through 90°. The units shall be capable of following a bend radius of 500 mm.

(f) The method of fastening the unit to the wooden door shall have an adjustment allowance of ± 20 mm on the mean height position.

(g) Two units are to be fastened to each door, except the leading door which has only one unit fastened to it. (This is to allow for normal hinge closure.)

6. Performance

(a) Each unit shall be capable of supporting a load of 250 N.

(b) The maximum force necessary to slide the whole garage door assembly shall not exceed 60 N.

7. Life and Reliability

Each unit shall have a useful life of seven years when operated four times each day.

8. Information and After-sales Service

(a) RCT Ltd. shall be supplied with a leaflet giving fitting instructions for the unit. This leaflet shall be included in each package of units.

(b) Brokham Bros shall operate an after-sales service to deal with repairs and replacements where necessary.

Synthesis, Comparison and Evaluation

Fig. 6.2 shows some of the alternative methods which could be considered to give free motion to the unit. Fig. 6.2(a) utilises a flat surface coated with anti-friction material, e.g. a thin strip of PTFE or nylon. Fig. 6.2(b) makes use of a ball and Fig. 6.2(c) utilises a roller. In method (a), while the coating of anti-friction material would reduce the friction between the unit and the track, it may still result in a heavy action and may be difficult to incorporate in a design which must follow a radiused track. Method (b) would certainly reduce the friction between the unit and the track but the resulting point contact, between mating surfaces, could lead to unacceptable wear of the track. Method (c) would similarly have reduced friction between mating surfaces and would have the advantage over method (b) of line contact between the unit and the track, thus spreading the load and reducing the wear risk. At this stage method (c) looks to be the most satisfactory basis for further consideration.

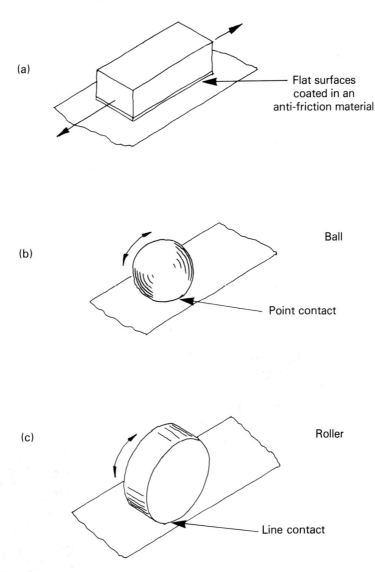

Fig. 6.2 Simple sketches showing alternative means of providing free-running motion to a sliding door unit.

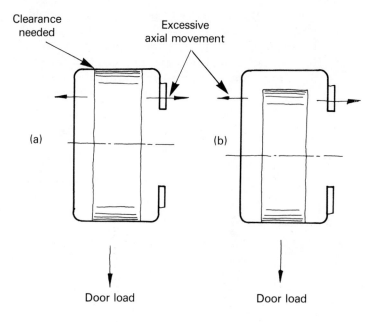

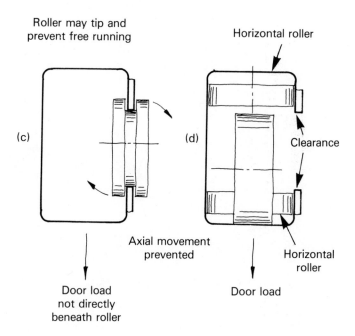

Fig. 6.3 Sketches showing possible roller configurations (internal outline of track shown only).

Let us give further thought to the use of rollers. Fig. 6.3(a) shows a roller 50 mm diameter which would locate on the top and bottom inner faces of the track. This method could be ideal, in theory, for it would limit any lateral rocking movement of the unit. However, a small clearance would be needed between roller face and track, for whenever contact is made with track faces simultaneously the system would jam. This would also render the system susceptible to variations in track size. Method (b) shows a smaller roller which consequently alleviates the risk of jamming. Both these methods have the advantage of the door weight being immediately in line with the action of the roller. A disadvantage is the excessive axial movement which both designs permit. Method (c) utilises a double-flanged roller running on the edge of the track section. The weight of the door does not act in the line of the roller action and would result in a rocking action of the door, which could prevent free running. The width between roller flanges may be designed to limit this rocking action, but sufficient clearance must be allowed here to permit movement along the radiused part of the track. Axial movement is prevented in this design. Method (d) utilises a more comprehensive system to restrict axial movement and still allows the weight of the door to act in the line of the rollers' action. The same problem arises as with method (a), however, for a slight clearance must be permitted between the horizontal rollers and the inner track faces, as simultaneous contact would cause jamming.

Bearing in mind the advantages and disadvantages of the four possibilities, shown in Fig. 6.3, let us consider the number of rollers per system. Fig. 6.4(a) shows two rollers per door (one roller per unit) and Fig. 6.4(b) shows four rollers per door (two rollers per unit). Fig. 6.5 shows pictorial views of these possibilities in single and tandem roller assemblies.

The configuration shown in Fig. 6.5(d) may be the most suitable, but further problems could arise when consideration is given to the fastening of four rollers in such close proximity to each other. Let us follow this line of thought and see if this system is viable. Fig. 6.6 shows a possible solution for retaining the rollers using two mild steel pressings.

Fig. 6.7 shows another possible solution using a two-roller configuration where axial movement is reduced by increasing the width of the rollers so that their chamfered corners bear upon the inside radius of the track. This system presents another problem. Because the width of roller has been increased to almost the inside width of the track the method of roller retainment becomes more difficult. The roller would have to be counterbored in order to accommodate the fastenings.

Both the designs shown in Fig. 6.6 and Fig. 6.7 would be quite expensive to manufacture and result in an unacceptable selling price.

In view of the relatively light load on the unit (250 N, see specification), tandem rollers would not be necessary from a pure loading point of view. They may increase stability, but the cost of extra rollers, together with their retaining systems, could not really be justified.

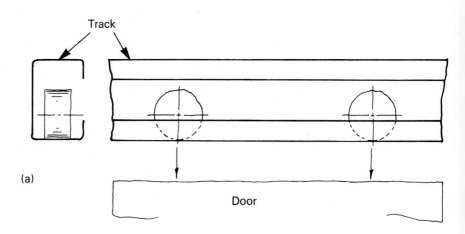

(a)

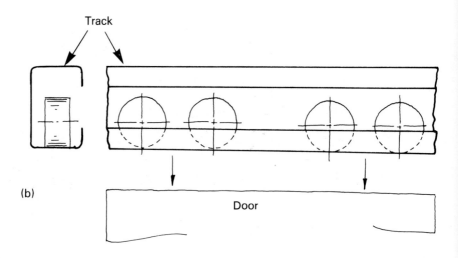

(b)

Fig. 6.4   Sketches showing a sliding door system with   (a) two rollers.   (b) four rollers per door.

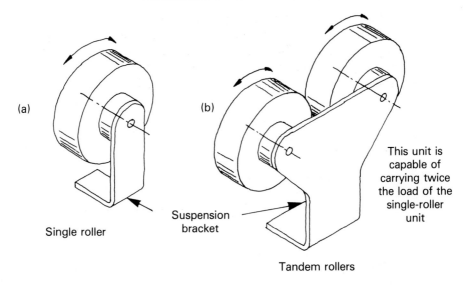

(a)

(b)

This unit is capable of carrying twice the load of the single-roller unit

Single roller

Suspension bracket

Tandem rollers

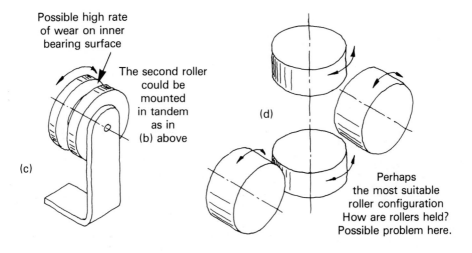

Possible high rate of wear on inner bearing surface

The second roller could be mounted in tandem as in (b) above

(d)

(c)

Perhaps the most suitable roller configuration How are rollers held? Possible problem here.

Fig. 6.5 Pictorial sketches of roller configurations corresponding to Fig. 6.3 (a) – (d).

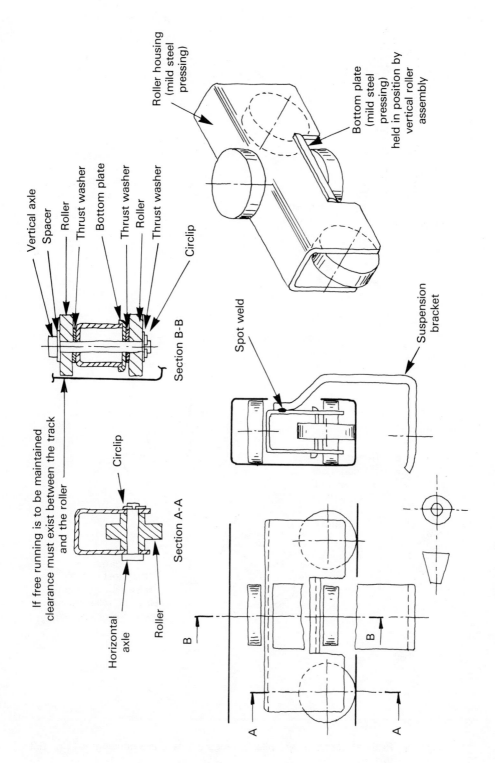

Roller housing (mild steel pressing)

Bottom plate (mild steel pressing) held in position by vertical roller assembly

Vertical axle
Spacer
Roller
Thrust washer
Bottom plate
Thrust washer
Roller
Thrust washer
Circlip

If free running is to be maintained clearance must exist between the track and the roller

Section B-B

Spot weld

Suspension bracket

Horizontal axle
Roller
Circlip

Section A-A

B

B

A

A

Fig. 6.6   A possible solution, in sketch form, for retaining rollers, in the four-roller configuration of Fig. 6.5 (d) using two mild steel pressings.

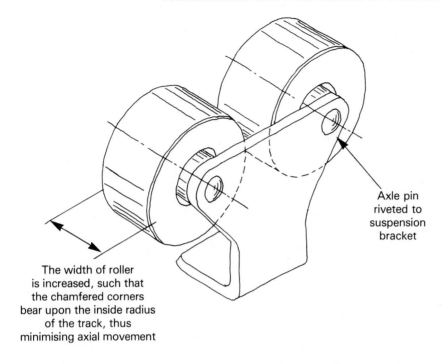

The width of roller
is increased, such that
the chamfered corners
bear upon the inside radius
of the track, thus
minimising axial movement

Axle pin
riveted to
suspension
bracket

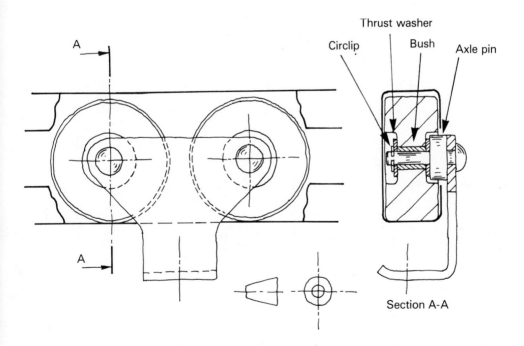

Thrust washer

Circlip   Bush   Axle pin

A

A

Section A-A

Fig. 6.7   A possible solution, in sketch form, to the two-roller configuration of Fig. 6.5 (b).

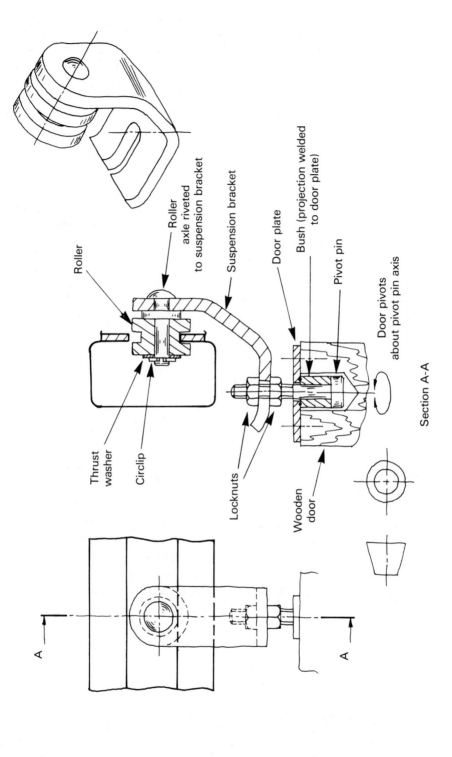

Roller

Roller
axle riveted
to suspension bracket

Suspension bracket

Door plate

Bush (projection welded
to door plate)

Pivot pin

Door pivots
about pivot pin axis

Section A-A

Thrust
washer

Circlip

Locknuts

Wooden
door

A

A

Fig. 6.8 A possible solution, in sketch form, to the roller configuration shown in Fig. 6.3 (c).

Let us examine a system, which could be developed, from the idea shown in Fig. 6.3(c) — the double-flanged roller design. One of its advantages is the prevention of axial movement. Its main disadvantage is the likely rocking action from an offset door loading. If we consider introducing a bottom door guide, this disadvantage is eliminated. Fig. 6.8 shows this system in orthographic projection and also in pictorial view. A double-flanged roller, made of brass, would be mounted on to a mild steel axle and secured by a circlip. The axle would be fixed to the suspension bracket by riveting. A door plate and bush would secure the pivot pin to the wooden door in order that it could pivot about its own axis (pivoting is necessary to allow the door smooth movement around the radiused section of the track). The pivot pin would be fastened to the suspension bracket by two locknuts to allow vertical adjustment. Fig. 6.9 shows the bottom door guide in pictorial view.

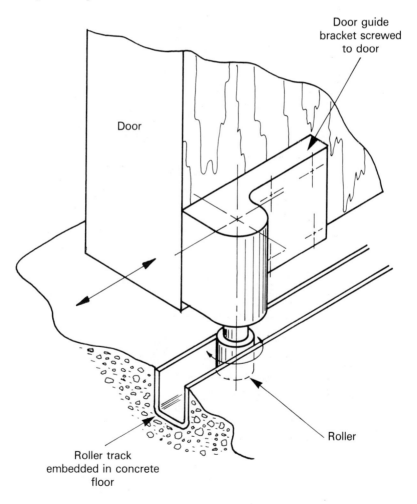

Fig. 6.9   Details of a bottom door guide.

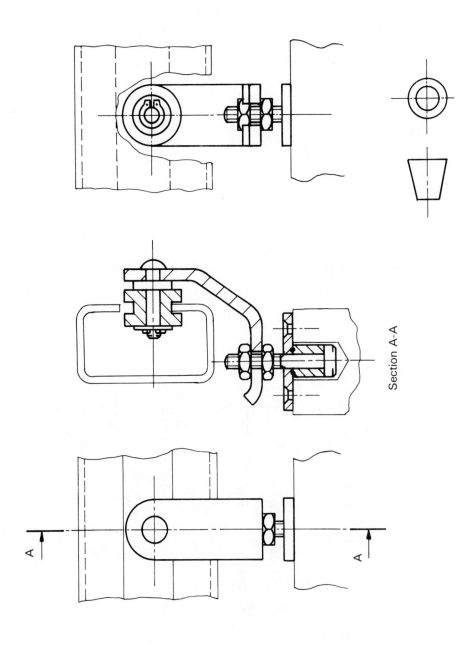

Section A-A

Fig. 6.10 An assembly drawing of the sliding door suspension unit.

The design team acknowledge that this design is the most practical and will be economic to produce based on the requirements of the job. They invite the Production Services Department to examine the design to see if there are likely to be any difficulties in manufacture which they may have overlooked. The answer is no, and the designer then decides to proceed with the production of the Assembly drawing (see Fig. 6.10) and the Detail drawings. This design together with a price was to be submitted to RCT Ltd.

Calculations

1. Based on a sense of proportion the pivot pin diameter should be about 6 mm.

    Let us check the factor of safety to see if this diameter is satisfactory.

    Maximum force on pivot pin $= 250\,\text{N}$ tensile   (see specification, item 6(a))

    Maximum tensile stress of mild steel $= 470\,\text{MN/m}^2$

    Cross-sectional area of pivot pin at root diameter (say 4.9 mm)

    $$= \frac{\pi d^2}{4} = \frac{\pi \times 4.9^2}{4} = 18.9\,\text{mm}^2$$

    $$\therefore \text{ Working stress} = \frac{\text{Force}}{\text{Cross-sectional area}} = \frac{250\,\text{N}}{18.9\,\text{mm}^2}$$

    $$= 13.2\,\text{N/mm}^2$$

    $$= 13.2\,\text{MN/m}^2$$

    $$\therefore \text{ Factor of safety} = \frac{\text{Maximum stress}}{\text{Working stress}} = \frac{470\,\text{MN/m}^2}{13.2\,\text{MN/m}^2} = 36 \text{ approx}$$

    Reference to Table 1.5 shows that a general factor of safety of 12 would be used for steel components subject to live loads with shock. Consequently a pivot pin 6 mm diameter is more than amply strong enough in this application.

2. Based on a sense of proportion the axle diameter should be about 6 mm also.

    Let us again check the factor of safety to see if this diameter is satisfactory.

The axle is in single shear and the maximum shear load $= 250\,\text{N}$.

$$\text{Shear stress of mild steel} = 375\,\text{MN/m}^2$$

$$\text{Cross-sectional area of axle} = \frac{\pi d^2}{4} = \frac{\pi \times 6^2}{4} = 28.3\,\text{mm}^2$$

$$\therefore \text{Working shear stress} = \frac{\text{Force}}{\text{Cross-sectional area resisting shear}}$$

$$= \frac{250}{28.3} = 8.8\,\text{N/mm}^2 = 8.8\,\text{MN/m}^2$$

$$\therefore \text{Factor of safety} = \frac{\text{Maximum stress}}{\text{Working stress}} = \frac{375\,\text{MN/m}^2}{8.8\,\text{MN/m}^2} = 43\,\text{approx}$$

As with the first calculation, the choice of diameter, based on a sense of proportion, shows that 6 mm is more than adequate to do the job.

THE NEED FOR REDESIGN

During the service life of a product occasions may arise when it may have to undergo a redesign. The factors that may influence this are many and varied. Let us consider some of them.

(i) The appearance of a product may need improving to maintain current fashions and trends. The reader has only to contemplate how motor car styling changes periodically, for example the 'hatch-back' trend. It must not be forgotten, however, that outside influences, other than fashion, may dictate styling changes. In the automobile industry, for example, a higher maximum speed, improved fuel economy and legislation regarding safety, etc., will effect changes.

(ii) The product may not fulfil its intended purpose – its function – as expected. Perhaps, for example, the strength and weight characteristics of the material used for the product may require improvement. This may be achieved by using an alternative material.

Considerable advances are being made in materials technology, and the range of materials, particularly non-metallic materials such as plastics, is very wide. The trend towards usage of these materials, some of which have a high strength coupled with a low density, e.g. carbon fibre, is increasing.

It must be noted, however, that if a product is redesigned using a different material an alternative manufacturing process may have to be used.

Functional improvements in the product may be achieved by redesigning it in terms of shape. Ergonomics may also have to be considered here.

Another example of functional improvement is in machine tool drive systems where hydraulics and pneumatics are used to replace belt and gear transmissions.

(iii) An increase in material costs may necessitate the selection of an alternative cheaper material, if the product is still to be sold at a competitive price. However, in selecting an alternative material the designer should carefully consider the whole of the manufacturing economics, for an entirely different manufacturing process may have to be used, as all materials cannot be shaped in the same way. A different manufacturing process will also influence the design of the component, as we have seen in Chapter 2.

(iv) An expensive manufacturing process may have to be replaced by a more economical means of production, perhaps because of a drop in sales. Conversely an out-of-date manufacturing process may require replacement by a more modern process in order to obtain improved output. In either case the observations made in (iii), regarding material and product design, should be seriously considered.

Having considered a number of factors that may necessitate the redesign of an article, let us now examine a typical problem.

Six machines owned by a certain textile manufacturer underwent a recent overhaul, during which six grey cast iron transmission housing castings were found to be fractured, due to excessive wear. It was decided to replace them as they were beyond repair.

The company were informed by the machine makers that production of that particular machine ceased ten years earlier and replacement castings were no longer available. Because of the breakdown, production of fabric on the six machines had been halted. The machines needed to be put back into service urgently and the Chief Maintenance Engineer contacted a local steel and engineering contracting company in order that replacement castings could be made. Fig. 6.11 shows the the transmission housing casting together with eight critical dimensions which had to be maintained on the replacement housings.

The contracting designer suggested redesigning the housing as a welded steel fabrication and he based his evaluation on the following factors:

(i) As only six transmission housings were required, a welded fabrication would be much cheaper than a replacement

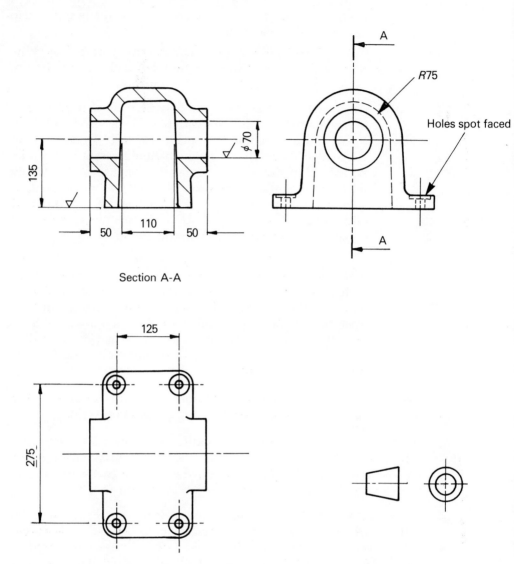

Fig. 6.11 Transmission housing casting, material: grey cast iron.

casting as the need for patterns, cores and mould preparation is eliminated. As a result a considerable improvement in delivery time would be achieved. These are important considerations as the textile company want to recommence their production as soon as possible with a minimum expense.

(ii) A welded fabrication is more sound. Problems such as porosity and shrinkage cavities, associated with casting, do not arise.

(iii) Because of the inherent strength deficiencies of grey cast iron and the nature of the sand-casting process (e.g. avoiding thin sections, which tend to allow the metal to cool quickly and hence permit it to solidify before it has completely filled the mould) castings produced in this material are usually made very robust and hence are often excessively heavy and bulky. Steel, on the other hand, has greater strength and because the operating conditions for the casting, or its fabricated replacement, are identical the dimensions of the fabrication can be reduced. For example, wall thicknesses can be reduced by as much as fifty per cent. Because of the reduction of dimensions there is also a considerable saving in weight.

A further advantage, not really relevant to this particular problem but nonetheless of considerable importance when evaluating the use of welded fabrications instead of castings, is that of being able to build up very large fabrications on-site, thus eliminating transportation, which may prove to be difficult or expensive with an alternative cast component.

The redesigned transmission housing is shown in Fig. 6.12. Note how the following points, together with some of those already mentioned, have affected the shape of the housing.

(i) Draft or taper has been eliminated.

(ii) Wall thicknesses have been reduced.

(iii) Machining of the base and spotfacing the holes in the base (which is necessary with a casting) has been eliminated.

(iv) The design allows access to the welds.

(v) Manufacture has been kept as simple as possible by using flat plates.

(vi) Welding has been kept to a minimum by using two plates bent at right angles to form the base and two vertical sides.

(vii) Critical dimensions have been maintained to facilitate assembly with existing components.

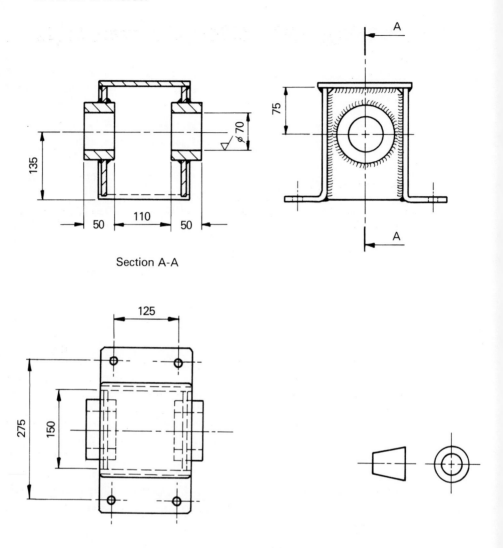

Section A-A

Fig. 6.12   Redesigned transmission housing using a mild-steel-welded fabrication.

Assignments in Design

After reading this chapter you should be able to:

★ carry out design assignments (G);

★ prepare assembly and detail drawings based on the design of three specific components:
e.g. a hand-operated clamp; a screw jack; a simple gear box; a hydraulic linear activator; a simple transfer mechanism (S).

(G) = general TEC objective
(S) = specific TEC objective

INTRODUCTION

The earlier chapters of this book are concerned with the choice of material for a product, the comparison of manufacturing processes, the ergonomic and safety aspects of design and the design of simple link and rotary mechanisms. In Chapter 6 we looked at the methods of preparing and evaluating designs and examples were examined in detail before arriving at a final design.

In this chapter you are required to carry out design assignments and prepare assembly and detail drawings based on the design of three of the specific components listed below. A specification is given for each component.

It is recommended that you follow the following procedure for the design of each component.

(i) Look at practical examples of the component. This enables you to get an appreciation of the product for you may not have an understanding of the working of such a component.

N.B. Do *not* think you can made a direct copy of the example, for this may infringe a Patent or Registered Design. In any case you are expected to contribute to an original design.

(ii) Make a number of designs of the component (freehand sketches will suffice).

(iii) Arrive at a final design by comparison and evaluation considering the quantity production and the consequent manufacturing processes available.

(iv) Prepare assembly and detail drawings based on the design, making stress calculations where appropriate to determine critical sizes.

A HAND-OPERATED CLAMP ————————————

Foreword

A simple hand-operated clamp is required to secure the component, during a welding operation, as illustrated in Fig. 7.1.

Definitions (Measuring Systems)

Metric units are used throughout. Masses are to be expressed in kilogrammes (kg) and forces are to be expressed in newtons (N). The Assembly and Detail drawings to be in accordance with BS.308:1972.

Related Documents

BS.5078:1974 Jig and fixture components.

Conditions

(i)   The clamp will be operated in a dust-laden welding shop.

(ii) Welding spatter may surround the clamp during operation.

(iii) The clamp will operate within a temperature range of 15–100°C.

Characteristics

(i)   The clamp must operate within the confines of the fixture base shown in Fig. 7.1.

(ii) The clamp must have a quick-release action.

(iii) The effort required to lock the clamp shall not exceed 50 N.

(iv) A quantity of one only is required.

Performance

The clamp shall be capable of exerting a force of 800 N on the workpiece.

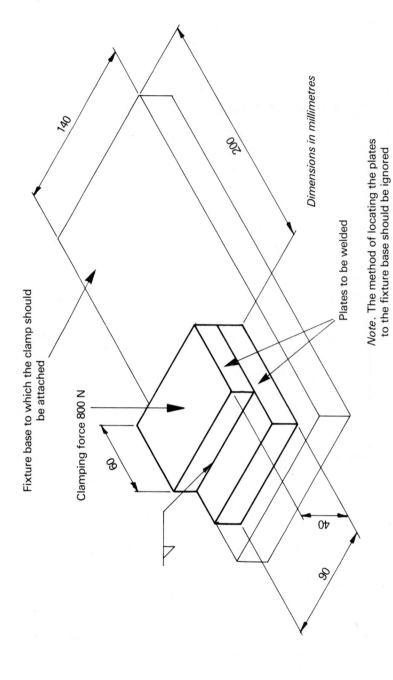

Fixture base to which the clamp should be attached

Clamping force 800 N

Plates to be welded

Dimensions in millimetres

Note. The method of locating the plates to the fixture base should be ignored

140

200

90

40

90

Fig. 7.1 Additional information for the hand-operated clamp.

A SCREW JACK

Foreword

A screw jack is required to help support a medium-sized steel casting during a milling operation.

Definitions (Measuring Systems)

Metric units are used throughout. Masses are to be expressed in kilogrammes (kg) and forces are to be expressed in newtons (N). The Assembly and Detail drawings to be in accordance with BS.308:1972.

Conditions

(i)   The head of the jack will be operating against an 'as-cast' surface.

(ii) The screw jack will operate in a workshop in a temperature range 15–30°C.

(iii) The screw jack will be splashed with soluble oil during its use.

(iv) The screw jack will be stored on an open shelf in the stores area.

(v)   Routine maintenance will be necessary to maintain easy mating between the screw and nut.

(vi) The screw jack will be subject to machining vibrations.

Characteristics

(i)   The screw jack shall be manufactured so that the head lies parallel to the base in any rotating position.

(ii) The screw jack shall operate from a minimum closed height of 75 mm to a maximum extended height of 125 mm.

(iii) The screw thread shall be of right-hand square form, and the screw shall advance a minimum of 2 mm and a maximum of 6 mm per revolution.

(iv) The mass of the screw jack shall not exceed 1.5 kg.

(v)   All parts of the screw jack that can come into contact with the hands of the user shall have no sharp edges.

(vi) The quantity production required is 10 only.

Performance

(i)   The screw jack shall support a load of 500 N.

(ii)   The effort required to lift the maximum load shall not exceed 20 N.

(iii) The efficiency of the system should be a minimum of 8% and a maximum of 30%.

A SIMPLE GEAR BOX

Foreword

A gear box is required for a special-purpose machine.

Definitions (Measuring Systems)

Metric units are used throughout. Masses are to be expressed in kilogrammes (kg) and forces are to be expressed in newtons (N). The Assembly and Detail drawings to be in accordance with BS.308:1972.

Related Documents

BS.2519 Part I:1976 *Glossary for gears.*

BS.3696 Part I:1977 *Specification for master gears.*

Conditions

(i)   The gear box will operate on a machine in a workshop and its working temperature range will be 15-50°C.

(ii)   The gears within the gear box will run in an oil bath.

Characteristics

(i)   The gear box shall be designed with a driving shaft, an idler shaft and a driven shaft.

(ii)   The driving shaft shall house a sliding cluster of two gears with 50 teeth and 25 teeth respectively.

(iii) The idler shaft shall house a fixed cluster of two gears.

(iv)   The driven shaft shall house a sliding cluster of two gears.

(v)   The overall size of the gear box shall not exceed 275 × 150 × 150 mm.

(vi)   The total mass of the gear box including gears, shafts and housing shall not exceed 5 kg.

(vii) Gear selectors are not required.

(viii) A quantity of one only is required.

(ix)   Spur gears should be used throughout.

(x)   The gears used should be in increments of five teeth. They should not be larger than 70 teeth and not smaller than 30 teeth.

Performance

    (i) The gear box shall have a range of four output speeds for a fixed input speed.

    (ii) The input speed (speed of driving shaft) shall be 882 rev/min.

    (iii) The four output speeds shall be approximately 184 rev/min, 490 rev/min, 630 rev/min and 1680 rev/min.

A conventional representation of gears is shown in Fig. 7.2.

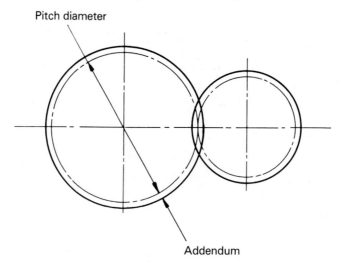

Let the addendum of each gear = 1.5 mm

The pitch diameter of a gear = Number of teeth on that gear × 1.5 mm

Fig. 7.2 Conventional representation of gears in the simple gear box.

A HYDRAULIC LINEAR ACTIVATOR ——————

Foreword

A hydraulic linear activator is required to operate a machine vice illustrated in Fig. 7.3.

Definitions (Measuring Systems)

Metric units are used throughout. Masses are to be expressed in kilogrammes (kg), forces to be expressed in newtons (N) and pressures to be expressed in meganewtons per square metre (MN/m^2). The Assembly and Detail drawings to be in accordance with BS.308:1972.

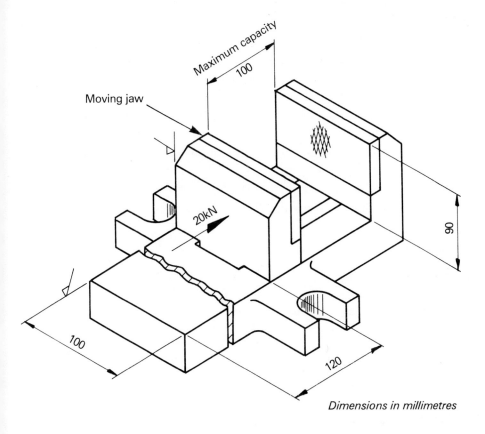

Maximum capacity
100

Moving jaw

20kN

90

100

120

Dimensions in millimetres

Fig. 7.3   Details of a machine vice to accept a hydraulic linear activator.

Related Documents

BS.4575 Part I: 1979 *Hydraulic Equipment*

Conditions

   (i) The activator will operate in a workshop in a temperature range 15–30°C.

  (ii) The activator will be splashed with soluble oil during its use.

 (iii) The activator and vice assembly will be stored on an open shelf in the stores area.

 (iv) The activator shall be maintenance-free.

  (v) The activator and machine assembly will be subject to machining vibrations.

Characteristics

(i) The activator shall operate in a position normal to the vice jaw movement.

(ii) The vice jaws shall operate from a fully closed position to 100 mm maximum open position.

(iii) The maximum clamping force exerted by the vice shall be 20 kN.

(iv) The diameter of the cylinder shall not exceed the height of the vice jaw.

(v) The pressure within the hydraulic circuit shall be 8 MN/m^2.

(vi) The activator shall be of double-acting design.

(vii) A quantity production of 50 units are required.

Performance

The maximum force exerted by the activator shall be equal to the maximum clamping force exerted by the vice.

A SIMPLE TRANSFER MECHANISM ————————

Foreword

A simple transfer mechanism is required to transfer damaged packets of cereals from a powered conveyor to a gravity roller conveyor initially at the same level. The general conveyor system is shown in Fig. 7.4.

Definitions (Measuring System)

Metric units are used throughout. Masses are to be expressed in kilogrammes (kg) and forces are to be expressed in newtons (N). The Assembly and Detail drawings to be in accordance with BS.308:1972.

Conditions

(i) The transfer mechanism will operate in a packing department in a temperature of $20 \pm 5°C$.

(ii) The transfer system will be foot-operated by female personnel.

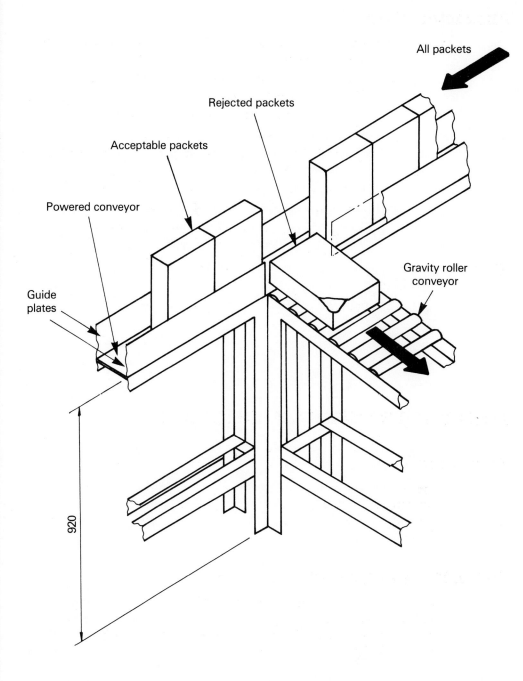

Fig. 7.4 A simple transfer mechanism.

Characteristics

(i) The working height of the powered conveyor and initial height of the gravity roller conveyor is 920 mm.

(ii) The maximum mass of each packet of cereal is 700 g.

(iii) The dimensions of the cereal packets are 320 mm high × 220 mm long × 80 mm wide.

(iv) The cereal packets travel down the powered conveyor, at a speed of 5 m/min, in an upright position with the 520 mm × 80 mm face leading.

(v) A manufacturing quantity of one mechanism is required.

(vi) As the mechanism is to be installed in a food preparation department, mineral oil lubricants must not be used. Consequently nylon-bearing materials should be incorporated.

Performance

The effort required to effect the transfer shall not exceed 25 N.

Further Reading

CHAPTER 1 ————————————————————————

W Alexander and A Street *Metals in the Service of Man* (Penguin, Harmonds-worth, 5th edn, reprinted 1973)

Controlling Corrosion, 1. Methods (Dept of Industry, Committee on Corrosion, London, 1976)

Handbook on Electroplating (W Canning, Spon, London, 22nd edn, 1979)

V B John *Introduction to Engineering Materials* (Macmillan, London, 1973)

M H A Kempster *Materials for Engineers* (Hodder and Stoughton Educ, London, 1975)

E Oberg and F D Jones *Machinery's Handbook* (Industrial Press, USA, 21st edn 1978. Distributed by Holt Saunders Ltd, Eastbourne)

CHAPTER 2 ————————————————————————

AWS Welding Handbook, Vol 2, Welding Processes (American Welding Society, 7th edn, 1978. Distributed by Macmillan Press, London)

P R Beeley *Foundry Technology* (Butterworth Group, London, 1972)

K J Clews *What goes on in Welding* (Rockweld Ltd, Woodhead Faulkner, Cambridge, 1974)

A C Davies *The Science and Practice of Welding* (Cambridge University Press, Cambridge, 7 edn, reprinted 1979)

Design Engineering Handbook — Metals (Morgan-Grampian Pub Co Ltd, London 1968)

Design Engineering Handbook — Plastics (Morgan-Grampian Pub Co Ltd, London 1970)

Die Casting, Yellow Back Series (The Machinery Publishing Co Ltd, London, 1970)

E Gregory and E Simons *Steel Working Processes* Odhams Books Ltd, London, 1964)

P T Houldcroft *Welding Processes (Engineering Design Guides 06)* (Oxford University Press, Oxford, 1975)

M H A Kempster *Materials for Engineers* (Hodder and Stoughton Educational, London, 1975)

W Kenyon *Basic Welding and Fabrication* (Pitman Publishing Ltd, London, 1979)

F Koenigsberger *Design for Welding in Mechanical Engineering* (Longman Group Ltd, Harlow, 1948)

R Matousek *Engineering Design* (Blackie and Son Ltd, 1963)

M F Spotts *Design Engineering Projects* (Prentice Hall, Inc, Englewood Cliffs, NJ, USA, 1968)

CHAPTER 3 ——————————————————————————————————

Ergonomics for Industry Booklets (Ministry of Technology, London)
  W T Singleton, *No 1, The Industrial Use of Ergonomics,* 1967
  B Shackel and D Whitfield, *No 2, Instruments and People,* 1967
  R G Sell, *No 5, Ergonomics v Accidents,* 1966
  K A Provins, *No 7, Man, Machines and Controls,* 1966

K F H Murrell *Ergonomics — Man in his Working Environment* (Chapman and Hall, London, 1979)

CHAPTER 4 ——————————————————————————————————

L A Beaufoy (ed) *Practical Mechanics for All* (Odhams Press Ltd, London, 1947)

British Standard BS.5034:1975, Code of Practice Safeguarding of Machinery (BSI, London, 1975)

Guide to Health and Safety at Work Act (British Safety Council, London, 1974)

W Handley (ed) *Industrial Safety Handbook* (McGraw-Hill Books Co (UK) Ltd, Maidenhead, 2nd edn, 1977)

Health and Safety at Work, etc. Act 1974 (HMSO, London, 1974)

Health and Safety Commission 'The Act Outlined' (Health and Safety at Work, etc, Act 1974 HSC 2, London, 1975)

V Powell-Smith *A Protection Handbook, Q and A on Health and Safety at Work Act* (Alan Osbourne and Associates Ltd, London, 2nd edn, revised 1975)

R G Sell *Ergonomics Versus Accidents, Ergonomics for Industry Booklet 5* (Ministry of Technology, London, 1966)

Answers to Numerical Questions

CHAPTER 1

1 132 mm

2 89.588 mm

3 313°C

4 5.16 mm³

5 87806 mm³

6 14°C

7 106 MN/m², 5

8 106 MN/m²

10 22 mm

11 27.5 kN, 280 MN/m²

22 4900